Collins

AQA GCS

Grade
Booster

Biology

Tom Adams

Contents

Introduction

About this Book

This book has been designed to support your preparation for the AQA GCSE Biology (9–1) examination and help you achieve your best possible grade.

The AQA Biology specification is divided into seven topics. This Grade Booster book mirrors that approach and the worked questions are divided into the same topics in the specification, so you can easily find questions that cover every topic you will study. As you revise each topic in this book, you will find exam-style questions, model answers and supporting notes with tips and hints. You will also find guidance on what the examiner is looking for and revision advice for different parts of the specification. Questions that only cover Higher Tier content are shown by this symbol: **HT**.

You can visit the AQA website to download or view a copy of the specification.

Terms in **bold** are defined in the Glossary at the back of the book. At the end of each chapter, you are signposted to pages in *Collins GCSE AQA Biology Revision Guide* (ISBN 9780008160678) for more information on the topics covered. The same page references apply to *Collins GCSE AQA Biology All-in-One Revision & Practice* (ISBN 9780008160746).

AQA GCSE Biology Exams

You will sit two exams, each of 1 hour 45 minutes duration.

In the final year of your GCSE course, your school will choose to enter you for either the Higher or Foundation Tier exam. If you are not sure which exam tier you have been entered for, talk to your biology teacher.

The information in the table below is the same whichever tier you sit.

	Paper 1	Paper 2
Topics covered	1. Cell biology 2. Organisation 3. Infection and response 4. Bioenergetics	5. Homeostasis and response 6. Inheritance, variation and evolution 7. Ecology
Exam marks	100	100
% of overall grade	50%	50%
Types of questions	Multiple choice Structured Closed short response Open response	Multiple choice Structured Closed short response Open response

Grading and Certification

The qualification will be graded on a nine-point scale: 1–9, where 9 is the highest grade.

If you are taking the Foundation Tier exam, then you will be awarded a grade within the range of 1–5. If you fail to reach the minimum standard for grade 1, you will be recorded as U (unclassified) and will not receive a qualification certificate.

If you are taking the Higher Tier exam, you will be awarded a grade within the range of 4–9. If you are sitting the Higher Tier exam and you narrowly fail to achieve grade 4, you will be awarded a grade 3. If you fail to reach the minimum standard for the allowed grade 3, you will be recorded as U (unclassified) and will not receive a qualification certificate.

Assessment Objectives

There are three assessment objectives (AOs) and the two exams will test these three different areas.

Assessment objective	Percentage of exam	Requirements
AO1 Demonstrate knowledge and understanding	40%	Demonstrate knowledge and understanding of: • scientific ideas • scientific techniques and procedures.
AO2 Apply knowledge and understanding	40%	Apply knowledge and understanding of: • scientific ideas • scientific enquiry • scientific techniques and procedures.
AO3 Analyse information and ideas	20%	Analyse information and ideas to: • interpret and evaluate • make judgements and draw conclusions • develop and improve experimental procedures.

Working Scientifically

The study of biology involves a lot of facts, theories and explanations but you will also get an understanding of the scientific process by thinking, discussing and reading about what scientists do. You will look at the contribution that some scientists have made to the world of biology and how ideas developed over time using the scientific process. You are also encouraged to appreciate how biologists communicate and check their ideas through scientific publications, and how ideas find practical and technological applications in everyday life.

Required Practical Activities

Biology is a practical subject. Many of the facts and ideas in this subject are derived from experiments, and theories can be proven or improved by experimental work.

The specification requires that you carry out ten required practical experiments, although you may well do more than this number of practical experiments during your course. Approximately 15% of marks in the exams will be based on the understanding that you have carried out these required practical activities. They will draw on the knowledge and understanding gained from having completed these practical activities.

There are many examples of questions based on the required practical activities throughout this book. You can find further information about the required practical activities in the AQA specification.

Mathematical Requirements

Throughout your study of biology you will need to understand a range of mathematical approaches to dealing with data. These range from simple calculations to use of statistics and converting units. Unlike Physics and Chemistry, there aren't a great deal of formulae and equations you need to recall but there are some! For example, make sure you have learned the photosynthesis and respiration equations.

You can find further information about the mathematical skills in the AQA specification.

Success in Biology

Biology as a subject can often be more descriptive than Chemistry and Physics, and the demands in terms of language and clarity of writing are high. For example, describing the processes of natural selection or how predator prey cycles work can be tricky. However, you will find that with practice you become more confident in expressing your ideas clearly in a way that gains credit in the exam.

Practising questions such as those given in this book will help you to assess your knowledge and understanding. It will also help you learn how to answer exam questions successfully. However, for factual information and to aid your understanding of the subject, use this book in conjunction with *Collins GCSE AQA Biology Revision Guide* (ISBN 9780008160678) or *Collins GCSE AQA Biology All-in-One Revision and Practice* (ISBN 9780008160746).

Command Words

Command words are specific words in questions that tell you what is expected in your answer. Command words can help you decide how to answer questions, how much detail to give and whether you are expected to simply recall information or give more explanation.

As you work through this book try to identify the command word in the question. If you are not sure of what the question is asking you to do then refer back to this list of words and explanations.

Calculate	use numbers given in the question to work out the answer
Choose	select from a range of alternatives
Compare	this requires the student to describe the similarities and/or differences between things, not just write about one
Complete	answers should be written in the space provided, for example on a diagram, in spaces in a sentence, or in a table
Define	specify the meaning of something
Describe	students may be asked to recall some facts, events or processes in an accurate way
Design	set out how something will be done
Determine	use given data or information to obtain an answer
Draw	produce, or add to, a diagram
Estimate	assign an approximate value
Evaluate	students should use the information supplied, as well as their knowledge and understanding, to consider evidence for and against
Explain	students should make something clear, or state the reasons for something happening
Give	only a short answer is required, not an explanation or a description
Identify	name or otherwise characterise
Justify	use evidence from the information supplied to support an answer
Label	provide appropriate names on a diagram
Measure	find an item of data for a given quantity

Name	only a short answer is required, not an explanation or a description; often can be answered with a single word, phrase or sentence
Plan	write a method
Plot	mark on a graph using data given
Predict	give a plausible outcome
Show	provide structured evidence to reach a conclusion
Sketch	draw approximately
Suggest	this term is used in questions where students need to apply their knowledge and understanding to a new situation
Use	the answer must be based on the information given in the question. Unless the information given in the question is used, no marks can be given. In some cases students might be asked to use their own knowledge and understanding
Write	only a short answer is required, not an explanation or a description

Animal and Plant Cells

When answering exam questions about different types of cells and the processes they go through, certain technical terms need to be learned and used correctly.

The following multiple-choice question illustrates how learning biological terms can result in achieving easy marks.

Example

Which of the following words best describes a bacterial cell? Tick **one** box.   *(1 mark)*

Prokaryotic   ☐

Multicellular   ☐

Eukaryotic   ☐

Undifferentiated   ☐

Prokaryotic ✓

You can quickly rule out 'multicellular' as, clearly, bacteria consist of one cell only (unicellular). More problematic is 'undifferentiated' but once you understand that the process of differentiation only occurs in multicellular organisms then this is also rejected for bacteria. The final two terms relate to the nature of cells themselves. Of the two, 'prokaryotic' is most appropriate because (among other things), prokaryotes have their DNA in the form of a loop (rather than a nucleus) and often contain plasmids. Of course, you could simply learn by rote that bacteria are prokaryotic, but if you have a thorough understanding of these terms you can apply the knowledge to more complex and extended questions.

Most students can recognise the differences between animal and plant cells from diagrams and pictures, but often the key differences are misunderstood and cause problems when students are presented with unusual animal and plant cells.

Example

This is a diagram showing two guard cells, which are found in the leaves of plants.

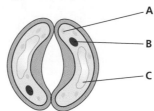

a) **i)** Name structure A. *(1 mark)*

A is a chloroplast ✓.

ii) Explain why this structure would not be found within an animal cell. *(2 marks)*

It would not be found in an animal cell because chloroplasts are sites of photosynthesis ✓, and animal cells do not photosynthesise ✓.

b) Which structure, B or C, would be found in almost all animal and plant cells? What is its function? *(2 marks)*

B (nucleus). It contains genetic material / chromosomes and controls other cell processes ✓ involved in cell division ✓.

A common type of question involves an understanding of scale and the ability to carry out simple magnification calculations.

Example

Darren is looking at some cheek cells under the microscope using high power. He clearly sees the nucleus, cytoplasm and cell membrane of the cheek cells. He wishes he could see mitochondria and ribosomes.

Give **two** reasons why he cannot see these structures. *(2 marks)*

✓ ✓ Any two from:

the organelles are too small to see
the microscope doesn't have a high enough resolving power
organelles need to be stained.

c) HT Using a specially fitted camera, Darren takes a picture of the cells he sees.

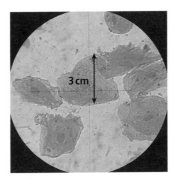

3 cm

Not to scale

He measures the diameter of one cell on his photograph by drawing a line and using a ruler. The line is shown on the picture. It measures 3 cm.

If the microscope magnifies the image 400 times, calculate the actual size of the cell in μm.

You can use the following formula:

$$\text{magnification} = \frac{\text{size of image}}{\text{size of real object}}$$

(3 marks)

$$\text{size of real object} = \frac{3}{400} \checkmark$$
$$= 0.0075 \text{ cm} \checkmark$$
$$= 75 \, \mu m \checkmark$$

Inserting the correct answer usually gains you full marks for this type of question, but you should always show your working because if you arrive at the wrong answer (perhaps by not pressing the correct key on your calculator or making a rounding error) then you can gain credit for showing how you intended to calculate the answer. To arrive at the answer you have to rearrange the formula to make 'size of real object' the subject of the formula,

i.e. size of real object = $\dfrac{\text{size of image}}{\text{magnification}}$

The example also includes an extra level of difficulty in that you have to show your answer as micrometres. You therefore need to know how many micrometres there are in a centimetre and convert accordingly. In this case, there are 10000 micrometres in a centimetre so multiply the answer (0.0075) by 10000.

Organisation

Levels of organisation within multicellular organisms refer to how cells are organised into tissues, tissues into organs and organs into organ systems. Simple exam questions are often aimed at testing your ability to appreciate order of magnitude and scale.

Example

Which of the following structures are in the correct size order, starting with the smallest? Tick **one** box.

(1 mark)

Tissue, organ, system, cell	☐	Cell, organ, system, tissue	☐
Cell, tissue, organ, system	☐	Tissue, system, cell, organ	☐

Cell, tissue, organ, system ✓

The correct order is usually quite easy to identify, but make sure you start with the correct item. For example, in this case, the question asks for smallest first. An easy mark can be lost by overlooking this detail.

Differentiation

Differentiation is related to cell specialisation in that multicellular organisms have a wide variety of different forms of cells, all arising from the same basic type. The process by which this is achieved is called differentiation.

Example

Diagram A shows a fertilised egg or **zygote**. All cells in the body arise from this one type, including cell B.

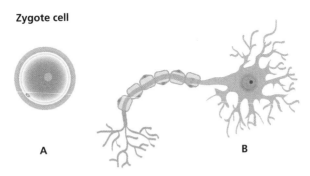

Zygote cell

A   B

a) Name cell B. *(1 mark)*

(Motor) *neurone / nerve cell* ✓

b) Explain how **one** adaptation of cell B allows it to perform its function efficiently. *(2 marks)*

Dendrites ✓ *allow connections to many other nerve cells* ✓.

Or

(Long) axon ✓ *allows transmission of nerve impulse over long distances* ✓.

Or

Myelin sheath ✓, *for insulation / to speed up nervous transmission* ✓.

To ensure that full credit is given, the proper technical name should be given to the structure (dendrites or axon) and full detail should be given to the adaptation (many other nerve cells / long distances). For example, the word 'many' is crucial in the example of the dendrites because a connection could still be made with only a single nerve fibre or axon. The point is that many connections allow control over responses through the synapses. They also allow a greater complexity (although these two reasons are not asked for in the question).

In the case of the 'axon' response, the emphasis here is on carrying the nerve impulse over a long distance – not just allowing the impulse to be carried.

Cell Biology

Stem cells refer to undifferentiated cells present in the embryo, some of which remain until adulthood. Stem cells have the potential to become almost any type of cell. Most exam questions focus on the use of these cells in medicine to treat a range of conditions.

Example

New research has taken place that allows a fully formed kidney to be developed from foetal cells obtained from amniotic fluid in the uterus (womb). In theory, doctors could collect amniotic fluid from around the baby, then store it for use in the baby's later life.

a) Describe how such cells could be useful in later life. *(2 marks)*

> ✓ ✓ Any two from:
>
> *for therapeutic cloning*
> *for producing cells with the same genes as the patient, e.g. for treating diabetes*
> Or
> *for transplant organs*
> *stem cells can be used to produce new organs that won't be rejected by the recipient, e.g. kidney.*

The important principles to stress here are that the stem cells can become any type of cell, and that they will have the same genes as the person in question.

b) Explain why the foetal cells are more useful than cells extracted from the individual several years after birth. *(2 marks)*

> *Cells in later life are already differentiated / become fixed ✓ and cannot become other types of cells ✓.*

c) Other uses of stem cells often require research using extracts from embryos. Explain **one** reason why some people might object to this type of research. *(2 marks)*

> ✓ ✓ Any two from:
>
> *an embryo is seen by some people as a living / sentient being*
> *and it would be unethical to experiment on its cells*
> Or
> *some stem cells might act as a reservoir of cancer cells*
> *that could then spread to other parts of the body.*

There are **two** marks for this question, so it is important to give **two** scientific points to gain full credit.

Culturing Microorganisms

There are three main ideas that may be tested in this section.

The first is the conditions needed for bacteria to grow.

Secondly, how **aseptic technique** can be used to grow pure colonies of microorganisms.

Thirdly, how bacterial growth can be assessed and measured.

Example

Shae has been given a pure sample of a bacterium *Staphylococcus aureus.* She wants to measure the numbers of bacterial cells in the culture and also investigate how their rate of growth is affected by certain factors.

Shae assesses the number of bacteria growing in the sample over a period of 8 hours and records her data. She plots data for the first 4 hours. The results are shown below.

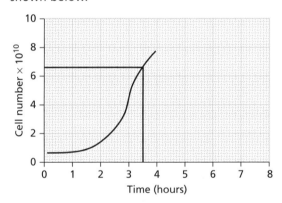

a) How many bacteria were present in the culture at 3.5 hours? *(1 mark)*

6.6×10^{10} ✓

Sometimes the correct units are supplied to you in the question. In this case, the answer is a number expressed in scientific notation, so the $\times 10^{10}$ part is important to add if the information is not given in the answer box. Also, take care to check scales. Here, each of the smallest divisions on the vertical scale represents 0.2. It is easy to assume that they are equivalent to 0.1 unless you check carefully.

b) At 4 hours Shae adds a 0.05% solution of disinfectant to the culture. On the graph, draw a line to show the likely growth curve between 4 and 8 hours. *(2 marks)*

> Almost any line would be acceptable as long as it:
>
> is not a steeper gradient than the original ✓
>
> continues to the 8-hour mark ✓.

So, the line could continue to rise at a lower gradient, it could level out or plateau, or it could decline / go down. (There is no knowing exactly how much bacteria would be affected, although the most likely result is that the line would start to go down as disinfectants kill bacteria.)

c) In order to carry out the next part of her experiment, Shae needs to grow several new cultures of the bacteria on separate nutrient agar plates. Describe how she would do this, giving the main precautions she would take to ensure that the colonies were pure, and how she would ensure that she achieved the correct conditions for growth. *(4 marks)*

> ✓ ✓ ✓ ✓ Any four from:
>
> transfer bacteria to agar using a sterilised inoculating loop, e.g. by passing it through a flame
>
> keep lid of petri dish open for the minimum time to avoid airborne microorganisms entering
>
> secure the lid of the petri dish with adhesive tape
>
> store plates / petri dishes upside down
>
> store cultures at 25°C in an oven.

You are not asked to explain why these steps are necessary. If you had been asked, reasons would have to be stated saying *why* these aspects of aseptic technique were used. Food, moisture and warmth are needed for the successful growth of bacteria, but the question has already stated that nutrient agar is being used, so only the mention of a correct temperature for incubation would gain credit. Note: temperatures of above 25°C are not used in school laboratories due to the risk of culturing harmful pathogens.

Questions on culturing bacteria are often combined with material from the 'infection and response' section of the AQA specification, so be prepared to apply your knowledge about disease and microorganisms.

Cell Division

Mitosis and **meiosis** are both forms of cell division and are often confused in terms of the process involved and the biological function of each one.

Example

Diagram A shows a vertical section through an onion root tip as seen under the light microscope at high power. Diagram B shows some of these cells at an even higher magnification.

A

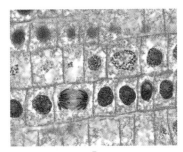

B

a)   Name the type of cell division taking place in the root tip.   *(1 mark)*

Mitosis ✓   The process can be identified because meiosis only occurs in reproductive tissue such as a testis or ovary. As an onion root tip is neither, the only option is mitosis. You might also deduce that root tips are rapidly growing meristems and growth is achieved by mitosis.

The graph below shows the masses of DNA found in the root tip cells.

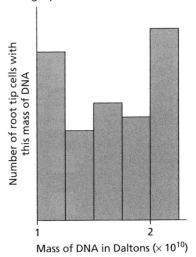

Number of root tip cells with this mass of DNA

1   2

Mass of DNA in Daltons ($\times 10^{10}$)

b) Explain why some cells have a mass of 1×10^{10} Daltons while others weigh 2×10^{10} Daltons. *(2 marks)*

> *Cells with the lower mass have the full chromosome number / have not had their DNA replicated yet (as part of mitosis)* ✓*. Cells with the higher mass have completed DNA replication / chromosome number has doubled (prior to the cell dividing)* ✓*.*

> You will be given credit for showing that you understand the process of mitosis: DNA is doubled up or replicated *then* the cell divides – which halves the mass again. The graph shows cells in both states as mitosis does not happen simultaneously between cells.

c) State another reason, apart from growth, why a cell might carry out this process of cell division. *(1 mark)*

> **Asexual reproduction** */ repair / replacement of old cells* ✓*.*

Example

Cells in reproductive organs divide by meiosis to form **gametes**. Describe what happens during this process and explain why changes in the number of chromosomes occur. *(4 marks)*

> This is an example of an extended answer question and examiners will be looking for **four** definite scientific points. You are likely to gain maximum credit if your sentences make sense when read together. Marks are often lost due to the explanation being difficult to follow – even though the student thinks they have correctly included the scientific information!

> ✓ ✓ ✓ Any three from:
>
> *copies of the genetic information are made*
>
> *the cell divides twice to form four gametes*
>
> *each daughter cell has a single set of chromosomes*
>
> *all gametes are genetically different from each other.*
>
> ✓ At least one explanation from:
>
> *gametes join at fertilisation / halve during meiosis to maintain the normal number of chromosomes*
>
> *meiosis / chromosome division allows* **variation** *to occur.*

Diffusion

There are three physical processes you need to be aware of that explain how materials enter and leave cells.

The first of these is diffusion. Questions will probe your understanding of the factors that affect the rate of diffusion, and adaptations found in organisms to increase this rate.

HT Example

Here is some data about a selection of different-sized organisms.

Organism	Surface area / m^2	Volume / m^3	Surface area (SA : vol ratio)
E. coli bacterium	6×10^{-12}	10^{-18}	
Malarial parasite	6×10^{-8}	10^{-12}	60 000:1
Honey bee	6×10^{-4}	10^{-6}	600:1
Roe deer	6×10^{0}	10^{0}	6:1
Sperm whale	6×10^{4}	10^{6}	0.06:1

a) Calculate the surface area : volume ratio of the *E. coli* bacterium. *(2 marks)*

6 000 000:1 ✓ ✓

To work out surface area : volume ratio, it seems a straightforward exercise at first to simply divide the surface area by the volume. However, plugging these values into a calculator will give you a number or decimal. Ratios can be tricky because of the way they are expressed.

There are many ways to tackle the exercise. Here's one: This problem is a little easier than most because you will notice that the ratios in the table are all multiples of 6. If we look at the two numbers, 6×10^{-12} and 10^{-18}, ignoring the scientific notation we see a simple relationship of 6 and 1. We just need to get the orders of magnitude right. If we subtract the indices (10^{-12} minus 10^{-18}) we get 10^{-6}. In other words, the surface area is one million times greater than the volume – hence 6 000 000 to 1.

One more thing – check the units of area and volume. In this case they are expressed in metres (squared and cubed), so we don't have to convert (thankfully – the question is difficult enough as it is – but then, it is Higher Tier!).

b) Using the data, explain why the sperm whale requires lungs and a circulation system to distribute oxygen to its tissues, whereas the *E. coli* bacterium does not. *(3 marks)*

Understanding SA : vol ratios requires you to do a 'flipping' exercise in your mind:

Large animal – small SA : vol ratio

Small animal – large SA : vol ratio

Notice that the question asks you to use the data, so you will not gain full credit unless you refer to the figures and compare them.

✓ ✓ Any two from:

large animals have small SA : vol ratios

passive diffusion is not efficient enough to deliver oxygen to such a large volume of tissue (in the sperm whale)

(therefore) a transport system is needed to transport the oxygen the long distances required.

Or any two from:

small animals have large SA : vol ratios

diffusion is efficient enough to deliver oxygen

because the diffusion path is short (in the bacterium).

And:

comparison of bacterium – 6 000 000:1 compared with sperm whale – 0.06:1 ✓.

c) Use ideas about surface area to explain how mesophyll cells in a plant leaf are adapted to maximise diffusion of water vapour to the environment. *(3 marks)*

Spongy mesophyll cells have air spaces between them ✓, which maximises their surface area ✓. Water evaporates from large surface areas more rapidly ✓.

You need to focus on the spongy mesophyll cells here as the palisade mesophyll cells are adapted for efficient photosynthesis, not evaporation of water / transpiration. Also, do not be drawn into referring to stomata. These control the amount of water vapour lost, which may mean minimising the surface area rather than maximising it.

Example

Joffrey is conducting an experiment to investigate the effect of surface area : volume ratio on the rate of diffusion.

- He takes some agar blocks containing a green-coloured indicator substance.
- Using a scalpel, he cuts out some jelly cubes with the dimensions shown in this table.

Dimensions of block / mm	SA : vol ratio (expressed as a decimal)	Time for block to completely change colour(s)
10 × 10 × 10	0.6	1400
10 × 10 × 5	0.8	1200
10 × 5 × 2	1.6	550
5 × 2 × 2	2.4	450
2 × 2 × 2	3.0	400

- He fills five boiling tubes with 1 M hydrochloric acid.
- He adds the blocks to the acid in the tubes and starts a stop-clock.
- The acid penetrates the blocks by diffusion and changes the colour of the indicator to red as it does so. Times are taken for each block to completely change colour.
- He records the results shown in the table.

a) Plot a graph of 'time taken for colour change' against 'SA : vol' on the grid. Join the points with an appropriate curve. The axes have been prepared for you.

(3 marks)

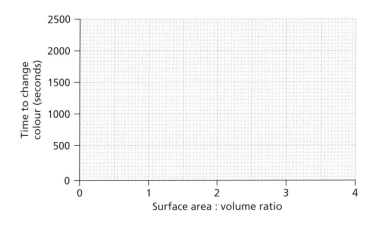

This question provides you with axes and scale – not all questions will. To gain full marks:

- Use a sharp pencil initially (you can rub out any mistakes!). Then go over the plots with a black pen. Circled dots or simple small crosses are acceptable.
- Check the scales – it's easy to misjudge what each small graph square represents. In this case it is:
  vertical (y-) axis: one small square represents 50 seconds
  horizontal (x-) axis: one small square represents 0.05.
- Use a pencil to produce a smooth curve then go over it in black ink. Be quite deliberate and try to avoid a shaky line (you will be given some leeway by the examiner but you will be penalised for a double line). The line should not extend beyond the data (i.e. past the first or last plots). The data in these questions will usually allow the line to pass through all the points unless there is an anomaly. Some questions ask you to identify an anomalous point (a point that does not fit the curve).

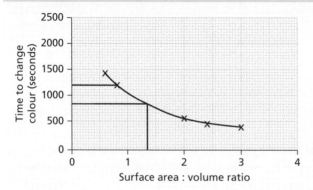

Correct plotting ✓ ✓
(subtract 1 mark for every incorrect plot)

Smooth curve ✓

b) What would be the SA : vol ratio of the block that would completely change colour at 800 seconds? *(1 mark)*

1.35 *(plus or minus 0.05)* ✓

Use a ruler to read off the scales (see answer graph in part a)). Again, understanding the scale is crucial. This question requires a high degree of precision (down to two decimal places) although there is a margin for error.

c) Describe the pattern shown by these results. *(2 marks)*

As the SA : vol ratio increases, the time for colour change decreases ✓. The effect decreases markedly after a SA : vol ratio of 2.0 / difference in time for increasing ratios is less ✓.

The first mark is always quite easy to pick up in these 'describing pattern' questions. The second mark is rarely gained because students fail to notice that this is not simply a proportional or inversely proportional relationship (straight line graph). If it was, you should state the straight line relationship.

d) Write down **one** way in which Joffrey could ensure that he obtained valid results in his investigation. Give a reason for your choice. *(2 marks)*

> ✓ ✓ Any two from:
>
> *ensure that all cubes are added at the same time*
> *so that each block has the same starting point / times can be fairly compared*
> *same volume of hydrochloric acid / agar blocks should be completely covered*
> *so that acid can penetrate all blocks equally*
> *temperature*
> *acid concentration.*

Validity is connected with 'fair testing' and you should concentrate on controlled variables to gain credit. Don't get tied up with accuracy or precision here, e.g. 'cut the blocks carefully'. There are many controlled variables you could choose – always look for the most obvious one(s).

Osmosis

The second passive process you are required to understand is **osmosis**, which is a special case of diffusion. Easy marks can be lost when explaining this process due to not defining terms accurately. Here's a classic definition of osmosis: 'The movement of water from a low solute concentration to a high solute concentration across a partially (or differentially) permeable membrane.'

There are various starting points to setting up your explanation of osmosis, and some will serve you better than others – in fact there is one that could be considered as the 'go to' choice.

Solute concentration – solute refers to the compound dissolved in the water. In biology examples this is usually salt or sugar. Notice that water moves against the solute concentration gradient – or low to high. It is this that confuses students because you are used to understanding diffusion occurring from high to low concentration. But remember, it is water that is moving and not the solute (and it certainly isn't the concentration that moves – as some candidates seem prone to describing!).

Water concentration – to avoid the above confusion, many students lay out their explanation in terms of water concentration as this allows them to equate it with the diffusion model of high to low that they are used to. This method is becoming frowned upon beyond GCSE and, although you will 'get away with it' at GCSE, it's best to opt for the next choice.

Cell Biology

Water potential – examiners recommend this approach because it is a measurable parameter that can be easily understood (and marked). To all intents and purposes, you can understand it in the same way as water concentration, but it has a more solid scientific basis. Perhaps it is best to think of it as a 'push' of water as if it was falling from a height.

The next example will illustrate this whilst looking at a particular application of the principle.

Example

Catelyn is investigating how plant cells respond to being surrounded by different concentrations of salt solution. She places rhubarb epidermal cells into 1.0 M salt solution and then observes them under the microscope. This is what she sees:

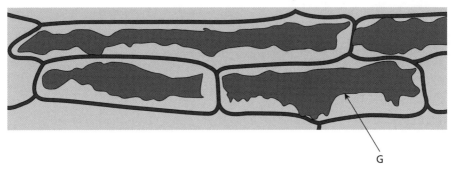

G

a) Name structure G. *(1 mark)*

Cytoplasm ✓

b) Describe and explain the appearance of the rhubarb cells. *(3 marks)*

✓ ✓ ✓ Any three from:

cytoplasm / protoplast has pulled away from the cell wall
this is **plasmolysis**
water has left the cells
by osmosis
water potential outside of cell is less than that inside / solute concentration is greater outside of cell / (water concentration is less outside of the cell)
water moves down the water potential gradient / against the solute concentration gradient.
turgor pressure on cell wall is reduced.

You see here that credit is given for using correct terminology like 'plasmolysed'. Always use the correct technical terms where possible. Also notice that credit is given for using a water concentration argument, but be aware that this might not be accepted in the future.

c) Catelyn removes the cells and places them in distilled water. Describe how the cells would change in appearance if she observed them under the microscope again. (*1 mark*)

✓ Accept any one from:

cytoplasm would be pushed up against cell wall / no gap between cell wall and cytoplasm

larger vacuole / vacuole swells or increases in size.

Active Transport

Unlike diffusion and osmosis, **active transport** requires energy to occur. You may be given situations in both animal and plant systems where active transport occurs. You need to be able to identify the features of active transport and apply them to the example. These features are:

* movement of particles / molecules / ions against a concentration gradient
* requires release of energy (from respiration).

Example

The diagram below shows a microscopic view of a root hair cell absorbing ions from the soil water that surrounds it. Explain how the ions can enter despite the fact that this is against a concentration gradient. (*2 marks*)

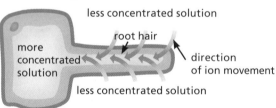

less concentrated solution

root hair

more concentrated solution

direction of ion movement

less concentrated solution

✓ ✓ Any two reasons from:

*ions enter by **active transport***

requiring release of energy

*from **respiration***

via protein carriers in the cell membrane.

Reference to protein carriers is no longer required for the AQA specification, but credit is usually given for this level of detail at GCSE. Once again, use of technical terms like 'active transport' gains easy marks.

For more on the topics covered in this chapter, see pages 8–15 of the *Collins GCSE AQA Biology Revision Guide*.

The Digestive System

You will have learned much about the anatomy of the digestive system at Key Stage 3 and will be required to have a thorough working knowledge of this for GCSE exams. Questions often focus on the connection between named organs and tissues, together with their role in both physical and chemical digestion.

Example

The diagram shows some organs in the human digestive system.

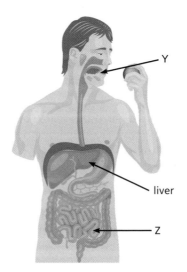

a) A process of absorption occurs in organ Z.

i) Name organ Z. *(1 mark)*

Small intestine / ileum ✓

ii) In order for food to be absorbed in organ Z, it has to be digested first. Explain why this has to occur. *(2 marks)*

To break large molecules down into smaller ones ✓, so they are small enough to pass across the gut wall ✓.

iii) The same type of enzyme (**amylase**) is produced in both organs Z and Y and acts on starch. Complete the word equation for this reaction.

starch   ___amylase___→ _____ *(1 mark)*

sugar / maltose ✓

AQA require you to know the enzyme groups: amylase, proteases and lipases. Glucose would be incorrect.

iv) A student carries out this reaction in a test tube using artificial amylase and a suspension of starch. State the reagent they would use to test for the product, and the practical steps needed to reveal a positive result. *(3 marks)*

This question relates to a required practical you will have carried out during your course. You will have used various reagents to test for a range of carbohydrates, lipids and proteins. It is important that you remember the names of these reagents, the experimental steps taken to use them and the result you would observe during a positive test.

Benedict's reagent ✓

✓ ✓ Any two from:

Add an equal volume / excess of reagent to the tube

heat in a water bath set at a temperature of at least 80°C

colour should change to green / orange / brick-red if reducing sugar (product of digestion) is present.

b) i) The liver is labelled on the diagram and is connected to an organ that stores bile. Name this organ. *(1 mark)*

Gall bladder ✓

ii) Bile emulsifies fat into droplets. Explain how this helps aid chemical digestion. *(2 marks)*

Increases the surface area of the lipid ✓*, which enables the enzyme lipase to attack / react with it more efficiently* ✓*.*

Many students can gain one mark on this type of 'explain' question, but few gain the full credit. When you see multiple marks available like this, ensure that you include sufficient detail / scientific points. It's worth checking this when you review your paper in the last ten minutes of the exam.

Enzymes

A number of common mistakes occur when students approach questions about enzymes.

- They are **molecules** (quite large ones), not cells and are not to be confused with hormones or antibodies.

- They are not alive and therefore they cannot be killed. You will lose marks if you make this mistake. They can, however, be **denatured**, which occurs when they are exposed to excessive heat or pH extremes.

- They break food molecules down without being chemically changed themselves.

Example

In an experiment to investigate the enzyme catalase, potato extract was added to a solution of hydrogen peroxide. The catalase in the potato catalysed the decomposition of the hydrogen peroxide and produced oxygen bubbles. The experiment was carried out at different temperatures and the results recorded in the table below.

Temperature (°C)	1	10	20	30	40	50	60	70	80
Number of bubbles produced in one minute	0	10	24	40	48	38	8	0	0

a) i)   Plot a graph of these results on graph paper and join the points with a smooth curve.

(3 marks)

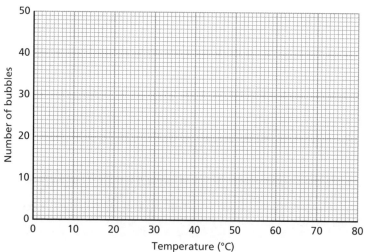

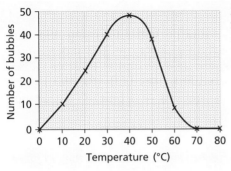

2 marks for correct plotting ✓ ✓
Subtract 1 mark for every incorrect plot.

1 mark for smooth curve ✓

ii) Describe how the number of bubbles produced varies with the temperature of the reacting mixture. *(2 marks)*

The rate of bubbles produced increases until it reaches an optimum / maximum ✓, then it decreases rapidly, producing no bubbles at 70°C ✓.

iii) Using the graph, estimate the optimum temperature for catalase to work at. *(1 mark)*

40°C ✓   The optimum temperature is obtained from reading off vertically from where the graph peaks. In this case the peak coincides with one of the plotted points but sometimes your curve needs to be extrapolated between two data points. As in all cases there will be a margin of error with the graph – usually equal to one small square (equivalent to 1°C in this example).

b) i) In 1890, Otto Fischer put forward a theory called 'lock and key', which provided a mechanism to explain how enzymes work. Explain, using lock and key theory, how catalase can break down hydrogen peroxide molecules. *(4 marks)*

✓ ✓ ✓ ✓ Any four from:
catalase / enzyme has a specific shape that will fit only the peroxide molecule (like a lock and key)
*this part of the enzyme molecule is called the **active site***
the enzyme puts a physical strain on the bonds in hydrogen peroxide
the peroxide molecule is broken apart
as the products no longer fit the active site.

ii) In 1958, Daniel Koshland updated Fischer's ideas and then published his findings. Apart from modifying existing theories, explain why scientists publish their work in scientific journals. *(1 mark)*

✓ Any one from:
to allow other scientists to test / verify their work
to share their ideas so they can be used for the benefit of all
to ensure that false claims are not made / allow peer review.

Breathing

You will need to recall the major organs and tissues within the human breathing system and be able to label these if provided with a diagram. It is important to distinguish between breathing (or ventilation) and respiration. Breathing is the process by which air enters the lungs using contractions of the diaphragm and intercostal muscles. Respiration refers to energy release in cells and is dealt with in the 'Bioenergetics' section.

Example

The lungs are organs of gaseous exchange. Explain how alveoli in the lungs are adapted to absorbing maximum oxygen into the bloodstream.   *(4 marks)*

Alveoli are present in large numbers and so provide a large surface area for the exchange of gases ✓, *the thin alveolar wall reduces the diffusion path* ✓, *the extensive capillary network maximises absorption of gases* ✓, *moist lining / epithelium of alveolus allows gases to dissolve / speeds up diffusion* ✓.

Take care in describing adaptations. You need to ensure that you link the feature to a reason and say why the structure represents a successful or efficient adaptation.

Questions about breathing are often connected to lung diseases or to smoking or cancer. Be prepared to relate your knowledge of the breathing system to data analysis.

Example

Scientists are studying the performance of pearl divers living on a Japanese island. They have taken measurements of the lungs of five 20–30 year-olds and timed how long they can stay underwater. The scientists also measured recovery time for the divers' breathing rates after a dive. The data is shown in the table below.

Vital capacity / litres	Max. time under water / mins	Time for breathing rate recovery / mins
3.5	2.5	3.2
4.0	2.7	2.9
4.3	2.8	3.5
4.5	2.9	2.8
4.6	3.0	2.5

a) Using your knowledge of the alveoli and diffusion, explain why pearl divers cannot stay underwater for longer than three minutes. *(2 marks)*

> ✓ ✓ Any two from:
>
> *concentration of oxygen in alveolus falls*
>
> *as it is not being replenished through breathing*
>
> *and oxygen is removed to the bloodstream*
>
> *concentration gradient falls.*

b) Vital capacity is the maximum volume of air that the lungs can contain. One of the team of scientists suggests that having a larger vital capacity allows a diver to stay underwater for longer. How well does the evidence support this conclusion? *(2 marks)*

> Agree (no mark): *because as vital capacity increased the time underwater also increased* ✓.
>
> Disagree (no mark): any one from: *only five subjects / sample size too small; need to include wider range of vital capacities* ✓.

> Notice that your decision to agree or disagree carries no marks as there are valid points on both sides of the argument. Also notice that to gain full marks you need to look at both possibilities and show an awareness of how data needs to be treated with caution. This is an element of working scientifically that appears throughout your exams.

c) The team decides that the vital capacity and breathing recovery rate is inconclusive. Suggest how they could obtain more valid data. *(2 marks)*

> ✓ ✓ Any two from:
>
> *test a wider range of subjects*
>
> *need to take account of the diver's gender*
>
> *standardise the ages (keep them the same)*
>
> *standardise the fitness levels.*

The Heart

Humans have a **double circulatory system** that allows blood to remain at a high pressure, as it needs to travel long distances around the body. This means that for every circuit of the body, the blood receives a boost two times – once after it has passed through the lungs, and once again when it has returned from the rest of the body. This idea is often tested in exams and could be coupled with questions regarding pressure changes in the heart.

Organisation

Example

This table shows pressure changes that occur in the left ventricle of the heart.

Time (s)	Pressure (kilopascals)
0.1	1
0.2	8
0.3	15
0.4	15
0.5	3

a) Describe the pattern shown by this data. *(2 marks)*

✓ ✓ Any two from:

pressure increases dramatically with time
until 0.3 / 0.4 seconds
then decreases (after 0.4 seconds).

Here is another example where full credit is gained for seeing all aspects of the pattern and identifying where changes occur – this is usually best achieved by quoting data.

b) At what time will most blood be present in the ventricle? *(1 mark)*

0.1 seconds ✓ Most blood will be present in the ventricle when it is at its most relaxed (ventricular diastole), which is just before contraction. In other words, when the pressure is least.

You need to be able to calculate blood flow rates. Interpreting this information and applying it to situations that might affect the heart are dealt with in the next example.

Example

Rate of blood flow can be calculated in a measure called **stroke volume**. This is the volume of blood pumped out of the heart every beat. The following formula is used:

$$\text{stroke volume (in litres)} = \frac{\text{volume of blood pumped by the heart per minute}}{\text{heart rate}}$$

a) Calculate the stroke volume of an athlete whose heart pumps 7.15 litres per minute at a heart rate of 65 beats per minute. *(2 marks)*

0.11 litres per beat ✓ ✓

If an incorrect answer is given, the correct working (7.15 ÷ 65) would still gain 1 mark.

This again illustrates the importance of showing full working in these calculations as credit can still be gained even if you arrive at the wrong answer.

b) In a condition called dilated cardiomyopathy, or DCM, stroke volume is lowered as the ventricles fail to contract properly. The diagrams below show a healthy heart and a heart with DCM.

Healthy heart   **DCM heart**

A

i) Name the chamber labelled **A** in the diagram.   *(1 mark)*

Right atrium / auricle ✓

ii) In the wall of this chamber there is specialised nervous tissue that keeps the heart beating in a controlled way. What is it called?   *(1 mark)*

Pacemaker ✓

iii) The data in the table below shows stroke volumes from healthy people and those with DCM.

	Normal heart	DCM heart
Stroke volume at 9 cm³	80	35

Using the information in the diagrams and the table, explain the effects of DCM on the structure and function of the heart, and suggest how this might affect the health of the patient.   *(3 marks)*

✓ ✓ Any two from:

(left) ventricle not able to fully contract

therefore less blood expelled from heart

stroke volume reduced.

✓ And one from:

more strain put on heart muscle

heart beats faster to compensate for lower stroke volume

symptoms might include shortness of breath, swelling of the legs, fatigue, weight gain, fainting, palpitations, dizziness, blood clots in the left ventricle (any one symptom).

Each part of the question needs to be addressed to obtain full marks. Biology questions often involve linking structure to function. In other words, describing a tissue, organ or system and then saying how the job of that structure is affected.

Blood Vessels

Arteries, veins and capillaries are the three main types of vessel you need to be aware of at GCSE.

Example

The diagram shows three kinds of blood vessel found in the human body.

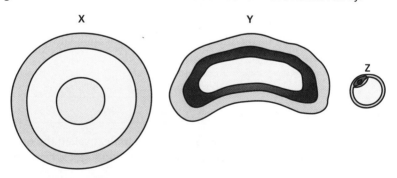

X   Y

Z

Diagram not to scale

a) Name vessels X and Y. *(2 marks)*

X: *artery* ✓

Y: *vein* ✓

b) Describe the differences between X and Y, and explain how these differences enable the vessels to carry out their function. *(3 marks)*

✓ ✓ ✓ Any three from:

X has narrower lumen / Y has wider lumen

wider lumen in Y gives less resistance to blood flow as pressure is less

X has thicker muscle / elastic wall

to withstand high pressure of blood / to even out pulses / allow smooth flow of blood.

> Notice that in comparison marks you need to be careful about wording (this has been mentioned before). For example, it would not be enough to say that Y has a wide lumen; use of the word 'wider' shows you are comparing it to something else.

c) Vessel Z is a capillary and has a diameter of 0.001 mm or 1 μm. Calculate the cross-sectional area of the vessel using this formula:

cross-sectional area = πr^2 *(2 marks)*

0.000000785 mm^2 or 0.785 μm^2 ✓ ✓

If an incorrect answer is given, correct working would still gain 1 mark.

> Radius is half the diameter, therefore 0.0005 mm. Using π as 3.14:
>
> 3.14 × (0.0005 × 0.0005) = 0.000000785 mm^2
>
> NB: the calculation is simpler if you use micrometre units.

The graph below shows the pressure changes that occur in the blood vessels lying between the aorta and the vena cava.

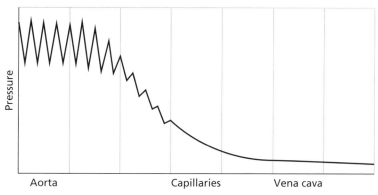

d) Explain why these changes in pressure occur. *(3 marks)*

✓ ✓ ✓ Any three from:

Aorta:
- high pressure
- caused by proximity to heart / muscular force of heart.

Capillaries:
- rapid / marked / severe drop in pressure
- due to resistance of capillaries with their small cross-sectional area.

Vena cava:
- pressure remains low as blood flow rate is slow / vessel is far away from / not yet received force from muscular contractions of heart.

Blood as a Tissue

Questions on blood as a tissue require you to learn the names of the various components and to relate their structures with their functions.

Example

The diagram below shows some of the components found in blood.

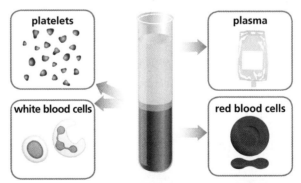

a) State **one** difference in the structure between red and white blood cells. *(1 mark)*

No nucleus in red blood cell / Hb versus no Hb ✓.

b) Why do red blood cells not possess a nucleus? *(1 mark)*

To allow room for molecules of oxygen to be carried / more Hb can fit in ✓.

c) With a diameter of 0.008 mm, a red blood cell is wider than the diameter of a capillary yet can still pass through. Explain how this is possible and why it aids the cell in carrying out its function. *(2 marks)*

Red blood cells are flexible / can fold / bend / squeeze through a capillary ✓.
Slower movement of the cell allows time for oxygen to diffuse out into tissues ✓.

The table shows the concentrations of some materials found in blood plasma.

Material	Concentration in millimoles per litre
Sodium	150
Calcium	0.8
Glucose	5
Protein	70

d) i) The average human body contains 3 L of plasma. How much glucose would this volume contain? (Give your answer in millimoles.) *(1 mark)*

15 mmol/L ✓

ii) This data was obtained from a healthy volunteer. Suggest how the concentrations of these solutes would differ in an untreated diabetic. Give a reason for your answer. *(2 marks)*

Glucose concentration higher ✓, due to lack of insulin / inability to absorb glucose from blood ✓.

This section of your course has been linked with ideas from the 'Homeostasis and response' section. This is a common thing to happen and you should practise linking knowledge between the topics for this reason.

Coronary Heart Disease

The AQA specification covers many non-communicable diseases and their treatments. It is impossible to give examples of every type of question you might encounter, but the following points are quite common:

- Comparing data in populations who suffer from heart disease. This may be shown in charts, graphs and tables. Viewing different presentations of this type of information will help you hone your interpretive skills.

- Describing the advantages and disadvantages of different treatments, for example heart valve replacements versus transplants or mechanical hearts.

- Discussing ethical and economic factors that affect how diseases are treated.

- Understanding the interactions between different factors. For example, heart disease may be affected by hereditary factors, drinking alcohol, diet, smoking, etc. and you should be aware that there is often no single factor that is solely responsible.

Example

a) The diagram below shows the stages of plaque build-up in a coronary artery (atherosclerosis). Such plaques can lead to life-threatening conditions. Use the information in the diagram to explain how plaques can lead to a heart attack. *(4 marks)*

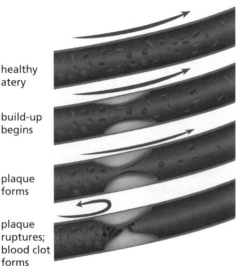

healthy atery

build-up begins

plaque forms

plaque ruptures; blood clot forms

✓ ✓ ✓ ✓ Any four from:

plaque restricts blood flow (in coronary artery)

blood clot more likely to form

oxygen cannot reach heart muscle

heart muscle not able to contract / ventricular fibrillation

heart muscle dies.

The third marking point is only gained if the word 'muscle' is used. Many students lose marks because they refer vaguely to the heart or ventricle.

b) One treatment for coronary heart disease is to introduce a **stent** to the affected artery. Describe how a stent can relieve the problems caused by plaque build-up.

(2 marks)

✓✓ Any two from:

(metal) tube / cylinder inserted into artery

cylinder expanded / inflated to widen artery / lumen

blood flows more freely

more oxygen can reach heart muscle.

c) Prevention of so-called 'lifestyle' diseases is a priority for health services as it saves money in the long term compared with treating the condition once it has happened.

Drugs called **statins** lower cholesterol and some have claimed that they prevent up to 80 000 heart attacks and strokes every year in the UK. The current advice given to GPs is that the benefits of statins far outweigh their risks.

However, there are possible side-effects. One review stated that there is a real risk of myopathy, a neuromuscular disorder that causes muscle damage. One in 10 000 people per year may develop myopathy as a result of taking statins. Another five to 10 people per 10 000 people will have a haemorrhagic stroke, which involves bleeding into the brain.

The graph below shows some data about how long people survive heart disease. It comes from a survey where one group did not take statins whereas the other did.

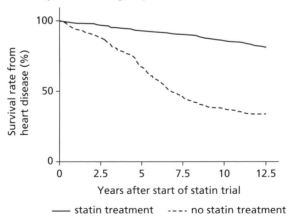

—— statin treatment   ---- no statin treatment

Using the information and the graph, say whether you think statins should be routinely prescribed to everyone over the age of 60 and give reasons for your decision. In your answer you should present arguments for and against prescribing statins.

(4 marks)

✓ ✓ ✓ ✓ Any four arguments from:

For:

statins increase survival rate / life expectancy

people are more likely to have heart disease when they are older (less value in giving statins to younger people)

risks of side-effects are sufficiently low / benefits of taking statins outweigh risks.

Against:

statins can cause life-threatening conditions

numbers may be low but they could be significant and other research needs to be conducted and studied

money could be better spent on other preventative means, e.g. advertising campaigns promoting healthy lifestyles

as healthier lifestyles are risk-free.

Smoking

Example

The table below shows data produced from a Stop Smoking campaign.

Age when men stopped smoking	Risk of cancer (%)
Non-smoker	0.4
30	1.0
40	2.8
50	5.6
60	11.0

a) Predict the risk of lung cancer if a man stopped smoking at 45 years of age. *(1 mark)*

Any answer in the range 2.9–5.5 % ✓.

b) Jorah says that all people who continue to smoke after 60 years should have to pay for NHS treatment arising from smoking-related diseases. Do you agree? Give reasons for your decision. *(2 marks)*

Marks can be gained from either response. ✓ ✓ Any two from the following:

Yes:

risk of having cancer greatly increases between the ages of 50 and 60

lung cancer risk is directly / causally connected to smoking, so the decision to carry on smoking places a burden on the health service / disease is avoidable.

No:

NHS treatment should be offered free whatever the cause of the health problem

difficult to prove that cancer is definitely related to smoking in an individual / cancer might be caused by something else

data only relates to men, and risks may be different for women

unfair to target one lifestyle disease / would have to be applied to alcohol and other drug-related conditions.

> Take care to be precise and clear when constructing answers to these types of questions. It is easy to fall into the trap of being vague and giving answers like 'It's their fault that they got the disease' or 'It's unfair to pick on smokers'.

c) Carcinogens in tar are partly responsible for lung cancer. Name **one** other harmful substance found in tobacco and explain how it affects the body.   *(2 marks)*

✓ ✓ Either:

carbon monoxide; combines irreversibly with haemoglobin

NB: no marks would be gained for the answer 'makes it harder to breathe'.

Or any one from:

nicotine / highly addictive / raises heart rate.

Example

Emphysema is a lung disease that increases the thickness of the surface of the lungs for gas exchange and reduces the total area available for gas exchange.

Two men did the same amount of exercise. One man was in good health and the other man had emphysema.

The results are shown in the table.

	Man A	Man B
Total air flowing into lungs (dm^3/min)	89.5	38.9
Oxygen entering blood (dm^3/min)	2.5	1.2

Organisation

a) Which man had more oxygen entering his blood? *(1 mark)*

Man A ✓

b) Explain why Man B struggled to carry out exercise. *(2 marks)*

✓ ✓ Any two from:

less oxygen absorbed into blood

longer diffusion pathway across alveolus wall

insufficient oxygen delivered to muscles to release energy for exercise.

Alcohol

Example

Healthy liver   **Cirrhosis**

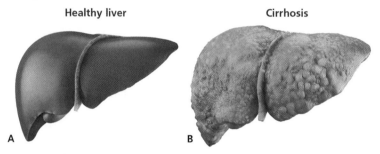

A   B

a) Look at the two pictures of the liver. The liver on the left (A) is healthy. The liver on the right (B) is from a person who suffered from a disease called cirrhosis.

Cirrhosis can be caused by excessive alcohol consumption. Explain how over-drinking can lead to cirrhosis of the liver. *(2 marks)*

Alcohol breakdown produces or releases toxins / poisons ✓.

Hardens / scars the liver tissue ✓.

You must state that it is the breakdown of the products of alcohol that cause the poisoning of the liver, not the alcohol itself. Note that no credit is given for saying that the liver becomes cirrhosed, as this is already stated in the question. You need to add extra detail to gain the second mark.

A study into alcohol purchasing habits was carried out in a European country. The chart below shows the average number of alcohol units purchased per heavy drinker per year in the study.

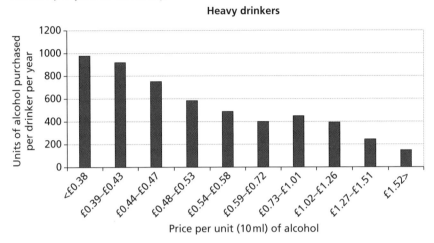

Heavy drinkers

b) Calculate the percentage decrease in units purchased between the unit price categories of £0.38 and £1.52. *(2 marks)*

81.6% ✓ ✓

If incorrect the following working can receive 1 mark: $\dfrac{980 - 180}{980} \times 100$

Where the reading of the graph values is slightly incorrect, answers will be accepted between 80% and 84% as long as they correspond to correct working.

The UK government is thinking of introducing a minimum alcohol unit price of 50p because they say irresponsible drinking costs the taxpayer £21 billion a year, with nearly a million alcohol-related violent crimes and 1.2 million alcohol-related hospital admissions. Others think that it will be unfair to responsible drinkers.

c) Discuss the advantages and disadvantages of introducing such a law. *(3 marks)*

This kind of question involves interpreting data, but also applying it to a situation where economic, health and other issues are involved. As stated many times before, the clarity of the points you make is important.

✓ ✓ ✓ Any three from:

increasing the unit price of alcohol generally reduces the units purchased this may lead to lower alcohol consumption / alcohol-related crime / deaths policy may damage alcoholic drinks industry / the economy / tax receipts may only affect the poor / have little effect on higher incomes / earners more alcohol may be purchased on the 'black market'.

Cancer

The main point to remember about cancer is that it occurs as a result of abnormal cell division. This is why it is connected with mitosis because normal cell division occurs by mitosis. You should also be aware that agents that cause cancer (carcinogens) are many and varied, although a popular group in exam questions are those contained in tar from cigarette smoke.

Example

Scientists investigating the effectiveness of an anti-cancer drug experimented on mice that had their lungs previously injected with a fast-acting carcinogen. One group was given a placebo as a control, while the other was given the anti-cancer drug. There were 10 mice in each group. After five weeks, the mice were killed and their lung tissue examined. The scientists counted the number of tumours they found. The results are shown in the table below.

Number of tumours	Number of mice	
	Group given placebo	Group given anti-cancer drug
0	0	6
1	0	2
2	3	1
3	2	0
4	2	1
5	2	0
6	1	0
7	0	0

The scientists concluded that the drug was effective and that there should be further trials.

a) One group was given a placebo – why? *(2 marks)*

A placebo is a medicine given to a subject without the active ingredient being present ✓, this is to ensure that any effects on the subjects (mice in this case) – beneficial or otherwise – are due to the anti-cancer drug and nothing else ✓.

b) What evidence is there that the drug is effective? Give a reason why the effectiveness is not necessarily reliable. *(2 marks)*

> Mice given the drug had fewer overall tumours / fewer tumours found on individual mice ✓.
> ✓ Any one from:
> but no account given to size of tumours
> sample size of 10 is quite small
> only one trial carried out.

c) Explain how a tumour could develop. *(2 marks)*

> ✓ ✓ Any two from:
> mutation in cell
> cell divides in an uncontrolled way / fault in the process of mitosis
> tumour formed from a group of these mutated cells.

d) The tumours in the mice were **malignant**. What does this mean? *(1 mark)*

> ✓ Any one from:
> the cancerous cells invade neighbouring tissues
> spread to other parts of the body
> secondary tumours formed.

e) The scientists did not recommend proceeding to market the drug immediately. Why not? *(2 marks)*

> ✓ ✓ Any two from:
> further research required to see if results are consistently produced
> other scientists need to verify / check the work
> use larger sample size of mice
> clinical trials on humans needed:
> to see if the drug is safe / to calculate an effective dose.

f) Some would say this experiment was unethical. Do you agree or disagree? Give a reason for your answer. *(1 mark)*

This type of question comes up often and you can give a generic answer based on the answers below most of the time. Just be careful not to simply say 'It's cruel'. You might get away with this type of response, but examiners are increasingly looking for greater clarity in these answers.

✓ Any acceptable reason.

Agree:

although the mice suffer / are killed, there is a greater good achieved through developing a drug that relieves suffering in humans / other animals, as the drugs tested on animals often end up being used by vets in the treatment of animals (idea of).

Disagree:

the mice have no choice as subjects in the experiment

there are other methods available for testing the drug / computer simulations / human volunteers.

Plant Tissues and Organs

Be sure to learn the technical names and functions of these plant tissues and organs. Spelling may be important as well, depending on the type of question. It's always good practice to spell correctly anyway.

Example

The diagram below shows a magnified view of the inside of a leaf. Complete the labels. *(3 marks)*

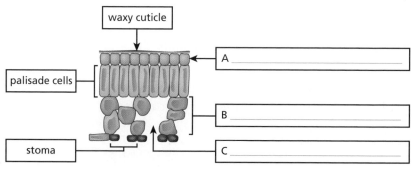

waxy cuticle

palisade cells

stoma

A _____

B _____

C _____

A – (upper) epidermis ✓ *B – spongy layer / mesophyll* ✓ *C – air space* ✓

Xylem and Phloem

Xylem and phloem are often confused in terms of their structure and function. As a quick comparison, always remember the following:

Part of plant	Appearance	Function	How they are adapted to their function
Xylem	Hollow tubes made from dead plant cells (the hollow centre is called a lumen)	Transport water and mineral ions from the roots to the rest of the plant in a process called transpiration	The cellulose cell walls are thickened and strengthened with a waterproof substance called lignin
Phloem	Columns of living cells	Translocate (move) cell sap containing sugars (particularly sucrose) from the leaves to the rest of the plant, where it is either used or stored	Phloem have pores in the end walls so that the cell sap can move from one phloem cell to the next

Example

The diagram shows the structure of phloem in the stem of a common plant. It also shows an aphid (not to scale) inserting its mouthparts into the phloem.

phloem

a) Explain why the aphid inserts its mouthparts into the phloem and not into the xylem. *(2 marks)*

Phloem carries dissolved sugars / sucrose ✓, which the aphid needs for energy / respiration ✓.

This question involves application of knowledge. You may not have heard of aphids before, but if you have revised the fact that the phloem carries sugar, then it isn't hard to deduce why the aphid's mouthparts might extract this rather than water from the xylem.

b) What name is given to the movement of substances around the phloem? *(1 mark)*

Translocation ✓

Organisation

c) Compare **two** structural differences between phloem and xylem tissue. *(2 marks)*

✓ ✓ Any two from:

xylem tubes are continuous hollow tubes / phloem have end walls / sieve tubes

xylem cells are not living / phloem cells are living

xylem cells are impregnated with lignin / phloem cells are not

phloem has companion cells / xylem does not.

> Take care to make your comparisons complete by saying what is present or absent in the phloem and xylem. For example, simply stating that xylem tubes are hollow might not be enough to get the mark if you don't state exactly how phloem cells differ. Some of the comparisons are outside the content required by the specification, but you would gain credit for them anyway.

Transpiration

Transpiration refers specifically to the evaporation of water from the surface of the plant's leaf. More specifically, the water evaporates from the film of water lying over spongy mesophyll cells within the leaf. This is why you need to have a thorough knowledge of plant tissues so that you can relate principles like transpiration to the appropriate structures.

You should also be aware that evaporation of water from leaves drives the whole transpiration stream. This is a continuous process from root to leaves. Many questions are based around the **potometer**, which measures transpiration rates under different conditions. This next example is slightly different but the same principles apply.

Example

In an experiment, a plant biologist carried out an investigation to measure the rate of transpiration in a privet shoot. She set up three tubes like the one in the diagram, measured their mass and exposed them each to different conditions.

A – left to stand in a rack

B – cold moving air from a fan was blown over it

C – a radiant heater was placed next to it

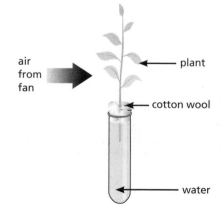

The tubes were left for six hours and then their masses were re-measured. The biologist recorded the masses in this table.

Tube	A	B	C
Mass at start (g)	41	43	45
Mass after six hours (g)	39	35	37
Mass loss (g)	2	8	5
% mass loss	4.87		11.9

a) Calculate the percentage mass loss in tube B. Show your working. Give your answer to three significant figures. *(2 marks)*

$\frac{8}{43}$ × 100 = 18.6%

✓ ✓ 2 marks for correct answer but, if incorrect, working gains 1 mark.

b) Which factor increased the rate of transpiration the most? *(1 mark)*

B (cold moving air) ✓

c) Evaporation from the leaves has increased in tubes B and C. Describe how this would affect water in the xylem vessels of the plant. *(1 mark)*

Water column in xylem would move upwards / towards the leaves more quickly ✓.

d) Explain how increasing the movement of air increases transpiration rate. *(2 marks)*

✓ ✓ Any two from:
air movement removes water vapour from air outside of leaf surface
diffusion gradient increased
diffusion of water molecules increased outwards.

Stomata and Root Hair Cells

Example

The diagrams on the next page show a magnified view of the lower leaf surface of a plant. The epidermal cells and stomata are clearly visible.

Organisation

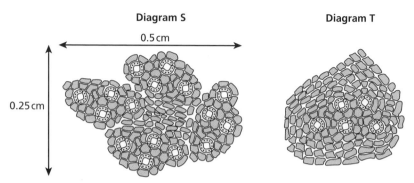

Diagram S

Diagram T

0.5 cm

0.25 cm

a) Calculate the stomatal density in the leaves of plant S. Express your answer in stomata per cm². *(3 marks)*

Area of leaf = 0.5 × 0.25 = 0.125 cm² ✓

16 stomata occupy this area, so the number of stomata occupying

$1 cm^2 = \frac{1}{0.125} × 16$ ✓

= 128 stomata per cm² ✓

> The correct answer on its own will yield three marks. But if this is incorrect, you will gain credit for each of the stages of working.

b) Plants lose water vapour through stomata. Explain how the position of stomata and changes in their structure can reduce water loss in plants. *(3 marks)*

Stomata on lower surface of leaf – means they are not as exposed to the heat of sunlight ✓.
Stomata can be closed – by action of guard cells to reduce aperture / exit route of water vapour ✓.
Sunken stomata – increases humidity around stomata and lowers the diffusion gradient ✓.

> Notice that the feature has to be linked to the scientific explanation to gain full credit.

c) Plant T has a stomatal density of 10 per cm². Suggest, with a reason, the habitat that plant T might be found in. *(2 marks)*

arid / dry / salt marsh ✓

✓ Any one of:

because lower numbers of stomata suggest the plants struggle to absorb enough water from the soil

fewer stomata means less water loss to counter the fact that little water is absorbed.

For more on the topics covered in this chapter, see pages 16–25 of the *Collins GCSE AQA Biology Revision Guide*.

Infection and Response

Communicable Diseases

Communicable diseases are those that can be spread via infection. You will need to revise the specific examples given in the specification and know what each one means, together with examples of diseases that are spread by those means.

The following question tests your knowledge of the four types of **pathogen** you need to know about, but you will also need to be aware of the differences between these microorganisms. Of particular note is the difference between bacteria and viruses in terms of size (viruses are much smaller) and the fact that viruses cannot survive for long outside of living cells.

Example

a) Link the type of microorganism to the disease it causes. (3 marks)

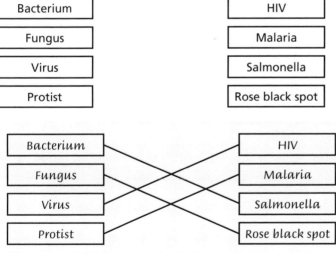

Bacterium	HIV
Fungus	Malaria
Virus	Salmonella
Protist	Rose black spot

All correct = ✓ ✓ ✓, three correct = ✓ ✓, one or two correct = ✓.

Infection and Response

b) The graph below shows the number of reported cases of salmonella per 100 000 of the population over the period of one year.

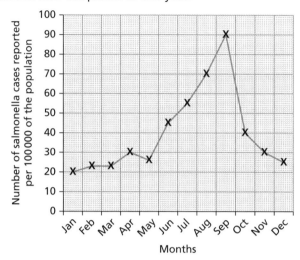

Months

Salmonella is spread when food is undercooked or stored at the wrong temperature. Use this information and the graph to describe and explain the trend shown. *(3 marks)*

✓ ✓ ✓ Any three from:

number of cases increase dramatically in Summer months / June to September e.g. from 44 per 100 000 to 88 per 100 000

warmer temperatures in Summer encourage growth of salmonella bacteria / vice-versa for Winter months

untended food warms up more rapidly in the summer months, encouraging bacterial growth.

Viral Infections

Example

John has caught measles and has been confined to bed for several days. John's mother was immunised against measles as a child via the MMR (measles, mumps and rubella) vaccine. She did not have John vaccinated as she was worried about reports in the media concerning a possible link between the MMR vaccine and autism.

a) How is John likely to have become infected with measles? *(1 mark)*

*Via **droplet infection** / virus carried in the air when a person with the disease coughs, sneezes or breathes ✓.*

b) State **one** symptom John may suffer as a result of the measles. *(1 mark)*

✓ Any one from:

fever   loss of sight

red skin rash   encephalitis

in some cases – fatal / death

c) The measles virus is a pathogen. What **two** ways do pathogens produce the symptoms that John experiences? *(2 marks)*

Cell damage ✓

Production of toxins ✓

d) The study that seemed to support a link between autism and the MMR vaccine was later discredited, even though it was originally published in a reputable scientific journal. The reputable journal later retracted the paper. Suggest how scientists might have discovered that the claim made about the link was false.

(1 mark)

✓ Any one from:

by studying the data again

by producing new evidence / looking at other studies

by checking the data analysis / conclusions drawn.

The other example from the specification of a viral infection you need to know about is HIV. This virus is sexually transmitted. It is important to realise that HIV refers to the virus, whereas AIDS is a description of the 'full-blown' symptoms resulting from untreated HIV infection. The following example brings in ideas about the immune system, which is typical of 'infection'-based questions.

Example

HT The photograph shows the structure of the human immunodeficiency virus (HIV). For decades HIV has spread throughout the world, especially in developing countries.

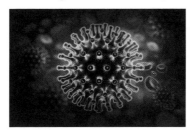

Infection and Response

A vaccine is now being developed that shows promising results. It works by mimicking the shapes and structures of HIV proteins. Scientists hope the immune system may be 'educated' to attack the real virus. A specially designed adenovirus shell can protect the vaccine genes until they are in a cell that can produce the vaccine protein.

Using your knowledge of the immune response and immunological memory, describe and explain how antibodies can be produced against the HIV virus. *(6 marks)*

✓ ✓ ✓ ✓ ✓ ✓ This is a model answer that would score the full 6 marks.

HIV proteins can be triggered and manufactured in existing human cells. The genes that code for the viral proteins are injected into the bloodstream. An adenovirus shell prevents them from being destroyed by the body's general defences. Once inside a cell, the genes instruct it to produce viral proteins that are presented at the cell surface membrane. The body's lymphocytes then recognise these antigens and produce antibodies against them. Memory cells sensitive to the viral proteins are stored in case the body is exposed to the antigens again.

This is an example of a 'level of response' type question. It is not marked in the usual way. Rather, the examiner will weigh up the scientific points you have made and how coherently you have presented them. This is why the answer is given in the form above instead of a series of marking points. Generally speaking, the levels are awarded as follows.

Level 3: A clear, logical and coherent answer, with no wasted words or excess detail. The student understands the process and links this to reasons for any appropriate experimental approaches. **5–6 marks**

Level 2: A partial answer with errors and ineffective reasoning or linkage. **3–4 marks**

Level 1: One or two relevant points but little linkage of points or logical reasoning. **1–2 marks**

Plant Infections

TMV and rose black spot are two pathogens the AQA specification requires you to understand. TMV is a viral pathogen that affects a variety of plants including tomatoes, while rose black spot obviously affects roses. This next example looks at both diseases and brings in ideas about disease control and the use of inorganic chemicals on the land.

Example

Rose black spot again. Fungicide is the only solution. It's straightforward and easy to apply.

It's better to avoid chemicals if you can – they're bad for the environment.

Dani   Ramsay

Dani and Ramsay both own rose gardens and want to produce high quality flowers for showing at competitions. Both their gardens have been affected by rose black spot in the past, but they differ in their approach to control.

a) Describe the symptoms and cause of rose black spot and discuss the options that Ramsay and Dani might consider. *(4 marks)*

> *Caused by a fungus ✓, which produces purple or black spots on leaves, which often turn yellow and drop early ✓. Treatment with fungicide – will kill the fungus on the leaves but disease may remain in soil or on other leaves / plants and cause re-infection ✓, removal of leaves / ensuring that plants are healthy / well-watered / have access to nutrients is more time-consuming but has a preventative effect too ✓.*

The mark scheme suggests how ideas can be expressed in an answer, but the examiner is interested in the idea itself, and as long as it is explained clearly you will gain credit.

b) Another plant pathogen is TMV. It can cause tomatoes to become stunted in their growth. Explain how TMV can have this effect. *(2 marks)*

> *TMV interferes with **photosynthesis** ✓. Photosynthesis produces carbohydrate for growth / reduced amount of sugar / starch / energy for growth ✓.*

Protist Diseases: Malaria

Questions on malaria are common and centre around knowledge of parasites, vectors and hosts together with the malarial parasite's life cycle and how the disease can be controlled. There is a link between this section and the section on Sexual and Asexual Reproduction' (page 97), where the parasite's life cycle is described. Take care to learn the terminology and practise applying knowledge of the life cycle, as this next example shows.

Example

Malaria kills many thousands of people every year. The disease is common in areas that have warm temperatures and stagnant water.

a) Explain why malaria is found in these areas.   *(2 marks)*

> ✓ ✓ Any two from:
>
> *malaria is transmitted by mosquitos*
>
> *warm temperatures are ideal for mosquitos to thrive*
>
> *stagnant water is an ideal habitat for mosquito eggs to be laid / larvae to survive.*

b) A protist called *Plasmodium* lives in the salivary glands of the female *Anopheles* mosquito.

From the box below, choose a word that describes each organism.   *(2 marks)*

parasite	disease	symptom	vector	consumer	host

Mosquito: _____

Plasmodium: _____

> *Mosquito: vector ✓*
> *Plasmodium: parasite ✓*

c) Samit, an African villager, believes that having mosquito nets around the beds of family members and taking antiviral remedies will reduce their risk of catching malaria. Explain why he is only partially correct.   *(2 marks)*

> *Nets will deter mosquitoes / prevent bites / prevent transferral of plasmodium ✓ but antivirals are ineffective as plasmodium is a protist / not a virus ✓.*

Non-specific Defences

The body has two main systems of defence against infection: non-specific and specific. As the names suggest, non-specific mechanisms generally protect the body and are adapted to prevent pathogens entering living cells. Specific mechanisms are able to target very defined groups of pathogens and are connected with antibodies and their products.

a) Below are the names of some defence mechanisms that the body uses. Match each defence mechanism with the correct function. The first one has been done for you. *(3 marks)*

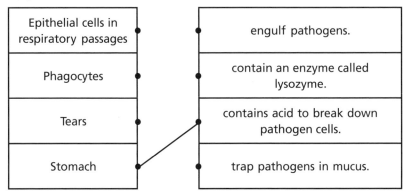

✓ for each correct line up to 3 marks. Subtract 1 mark for any additional lines.

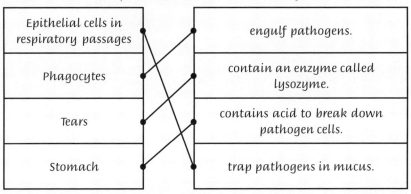

The main point to note here is not to add more lines than the three available. You will lose marks otherwise.

b) Epithelial cells in the trachea trap pathogens in mucus. But this does not get rid of them. Infected mucus can become a problem, causing bronchitis or even pneumonia. How do epithelial cells remove the trapped pathogens? *(2 marks)*

> *Beating **cilia** propel mucus and pathogens back up the trachea ✓, where they are then passed into the oesophagus / are swallowed ✓.*

Specific Defences and Vaccination

Make sure you understand the difference between the terms **antibody**, **antigen** and **antibiotic**. Here is a handy reference.

- Antibody – specific protein produced in response to a specific antigen on a pathogen. They are part of the body's defence system.

- Antigen – protein on the outside of a foreign cell / particle that can be recognised by antibodies.

- Antibiotic – medicine / drug produced to combat bacterial infections.

Example

The diagram shows a white blood cell producing small proteins as part of the body's immune system.

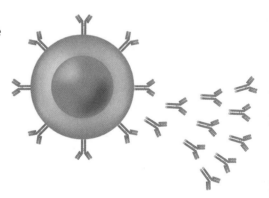

a) What is the name of these proteins? *(1 mark)*

> *Antibodies ✓*

b) These proteins will eventually lock on to specific invading pathogens. Describe what happens next to disable the pathogens. *(1 mark)*

> *Pathogens are clumped together to prevent their further reproduction, to make them easier for phagocytes to digest ✓.*

Example

The graph shows the antibody levels in a person after he contracted the flu. The flu pathogen first entered his body two days before point X. There was then a second invasion at point Y.

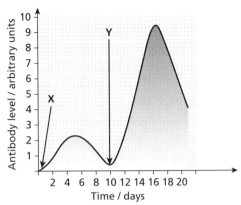

a) Name **one** transmission method by which Dominic could have caught the flu virus. *(1 mark)*

✓ Any one from:

droplet infection / coughing / sneezing / water droplets in breath / aerosol.

b) After how many days did the antibodies reach their maximum level? *(1 mark)*

Answer in the range 16–17 days ✓.

c) What is the difference in antibody level between point Y and this maximum? Show your working. *(2 marks)*

9 arbitrary units ✓ (9.5–0.5) ✓

d) Explain, using your knowledge of memory cells, the difference between these two levels. *(2 marks)*

Memory cells recognise future invasion of the pathogen ✓; they can produce the necessary antibodies much quicker, and at higher levels, if the same pathogen is detected again ✓ (i.e. when the secondary response occurs).

Vital extra marks are often lost by students because they do not grasp the idea of the body's response to vaccines and forget to mention **memory cells**. Remember, once antibodies have been produced against a particular antigen, the ability to respond quickly to that antigen on the cell surface membrane of a pathogen is built in to the body through memory cell formation. The massive response to a further infection ensures that the body does not suffer any symptoms. This whole process is triggered by vaccination, which is an artificial form of **active immunity**.

Infection and Response

Example

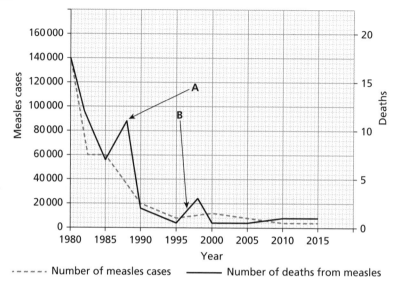

Number of measles cases ----- Number of deaths from measles

The graph above shows some data for a country's population between the years 1980 and 2015. Point **A** represents a time when a new vaccine, MMR, was introduced and most new babies received it. MMR stands for measles, mumps and rubella. Point **B** is when large numbers of people chose not to have their babies vaccinated using the MMR vaccine. Some did not have the vaccination at all, while others opted to have their children vaccinated against mumps, measles and rubella separately.

a) Describe the trend relating to the number of measles cases between 1980 and 1995.

(2 marks)

Measles cases fell overall between these dates ✓. There was a significant rise between 1985 and 1988 ✓.

b) Jan says that the introduction of the MMR vaccine accounts for the fall in cases of measles between 1980 and 1995. Explain the arguments for and against this conclusion.

(3 marks)

Cases fell significantly when the vaccine was introduced ✓, but cases had also fallen significantly between 1980 and 1985 / before the vaccine was introduced ✓ so other factors could have caused the drop in cases ✓.

c) In 1995 a paper was published that seemed to draw a link between the MMR vaccine and developing autism – a condition characterised by difficulty in communicating and forming relationships with other people. Evaluate the evidence in the graph that the publication of this paper caused an increase in deaths from measles. *(2 marks)*

> *Deaths / measles cases rose after this date* ✓; *this could be coincidence / no causal connection – simply correlation* ✓.

d) Within a year the findings of this paper were found to be suspect and there was a large increase in the uptake of the MMR vaccine again. Radesh says that parents who had their children vaccinated separately left them open to contracting measles. Do you agree? Give reasons for your answer. *(2 marks)*

> No (no marks)
>
> *There is no evidence from the graph that the rise in cases was due to the separate vaccine* ✓.
>
> *Rise in cases could be due to the parents who didn't have their children vaccinated at all / rise could be a coincidence* ✓.

e) Explain why declines in the uptake of a vaccine can be dangerous to the health of a population. *(2 marks)*

> *Loss of herd immunity* ✓.
>
> *More infected individuals are free to pass on pathogens to others* ✓.

f) What is contained within a vaccine? *(1 mark)*

> *A dead / weakened / heat-treated form of the pathogen* ✓.

Antibiotics and Painkillers

The topic of antibiotics reappears in topic 6 – Inheritance, Variation and Evolution. This is because bacteria are becoming increasingly resistant to antibiotics through the process of natural selection. So don't be surprised if this aspect is touched upon in paper 1 too.

The key thing to remember about antibiotics is that they do not work against viruses. Therefore, they should not be prescribed automatically by doctors as this increases the chance of resistance occurring (please note – resistance, not 'immunity' – bacteria do not have an immune system).

Another key point is that current health advice states that the full course of antibiotics should be taken if you are suffering from a bacterial infection.

Infection and Response

Example

Isoniazid was a drug developed in 1952 to treat tuberculosis (TB). Today one in seven new cases of TB is resistant to Isoniazid.

a) Explain as fully as you can **one** way in which this resistance could have arisen. *(4 marks)*

> *Within the population of TB bacteria there may be a few organisms with natural resistance to isoniazid* ✓. *This could be a result of a mutation* ✓. *When a patient is treated with isoniazid, all the sensitive bacteria are killed* ✓. *This allows resistant bacteria to quickly grow and multiply* ✓.

b) Nowadays it is common practice to treat patients with TB using two different antibiotics simultaneously. Explain how this can help reduce antibiotic-resistant strains emerging. *(2 marks)*

> *If a bacterium develops resistance to one of the antibiotics* ✓, *it will still be killed by the second antibiotic* ✓.

c) Doctors are concerned about the increase in the number of MRSA (methicillin-resistant *Staphylococcus aureus*) infections they are seeing. What can doctors do to reduce the likelihood of resistant strains emerging? *(2 marks)*

> *Not overprescribe antibiotics* ✓. *Not use antibiotics to treat minor infections* ✓.

Painkillers are used to treat the symptoms of disease but do not kill pathogens. In medical language, painkillers are called analgesics. Common painkillers are paracetamol, ibuprofen, aspirin and codeine. These can be obtained over the counter or on prescription. More potent painkillers exist such as morphine. These are used for patients who suffer extreme pain or who need end-of-life care. Questions on painkillers are rare but can form part of more general questions on drugs or clinical trials of drugs.

Discovery and Development of Drugs

Questions on clinical trials are common, and an example is given in this section, but the 'discovery' aspect is quite new. You are likely to be tested on factual recall regarding the discovery of drugs such as penicillin and the process of scientific experimentation and research.

Example

This is amazing. I left some cultures of bacteria to incubate on agar plates while I was on holiday. Upon my return, I found that a couple of the plates had become contaminated with mould colonies. Around the colonies there seemed to be an exclusion zone where no bacteria could grow. I think I may be on to something here.

Alexander Fleming

a) What further work would Fleming need to have done to prove that his discovery was not due to chance or some factor he had not been aware of at the time? *(2 marks)*

✓ ✓ Any two from:

repeat the experiment / conditions in his agar plates

record data / observations / measurements many times

set up suitable controls

publish work for others to check / verify

extend study using other strains of bacteria / mould.

b) In 1928, Alexander Fleming discovered the mould that made the drug which came to be known as penicillin. However, it wasn't until 1945 that mass production and distribution of the drug occurred.

Explain why there was such a long time period between discovery and production. *(3 marks)*

✓ ✓ ✓ Any three from:

drug / active ingredient needed to be extracted from the mould / clinical trials needed to be carried out:

to see if the drug was safe

to see if the drug was effective

to determine correct dosage.

Example

A pharmaceutical company is carrying out a clinical trial on a new drug called alketronol. They are testing it to see whether it produces significant adverse (harmful) events in a sample of 226 patients.

a) Apart from checking for adverse events, write down **two** other reasons why a company carries out clinical trials. (*2 marks*)

> ✓ ✓ Any two from:
>
> to ensure that the drug is actually effective / more effective than a placebo
>
> to work out the most effective dose / method of application
>
> to comply with legislation.

b) The kind of trial carried out is a double blind trial.
What does this term mean? (*2 marks*)

> Double blind trials involve volunteers who are randomly allocated to groups – neither they nor the doctors / scientists know if they have been given the new drug or a placebo ✓.
> This eliminates all bias from the test ✓.

c) Data from the trial is shown in the table below.

Adverse event	Number of patients	
	Taking Alketronol	Taking placebo
Pain	4	3
Cardiovascular	21	17
Dyspepsia	7	6
Rash	10	1

i) Calculate the percentage of patients in the trial who suffered a cardiovascular event while taking alketronol. Show your working. *(2 marks)*

Total patients = 226

$\dfrac{21}{226} \times 100$ ✓ = 9.29% ✓

ii) A scientist is worried that alketronol may trigger heart attacks. Is there evidence in the data to support this view?
Explain your answer. *(2 marks)*

Yes, there are more patients who took the drug and had a cardiovascular event than those who took the placebo. However, there is not a large difference between the groups ✓.

No, the cardiovascular events could include other conditions apart from heart attacks ✓.

iii) Which other adverse event shown in the table might cause concern? Give a reason for your decision. *(2 marks)*

Rash ✓.

The difference in numbers of patients who got a rash between the alketronol and placebo groups is quite large ✓.

🔴 Monoclonal Antibodies

Monoclonal antibodies are artificially produced in the laboratory to be used in the treatment of diseases and other technological applications. The process of producing these antibodies occurs through developing **hybridoma cells**. These cells divide rapidly (by cloning) and produce the required antibody. Millions of identical cloned cells are then made, which can be **purified** and used.

Once you get over the terminology involved then the process is not too difficult to understand as it relies on appreciating the immune process and antibody formation you have already studied. If you are still hazy on this aspect, then it is worth revisiting antibody production on pages 56–57.

Example

The diagram below shows how monoclonal antibodies can be produced in the laboratory.

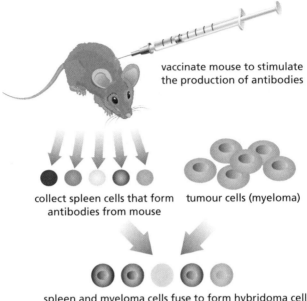

vaccinate mouse to stimulate the production of antibodies

collect spleen cells that form antibodies from mouse

tumour cells (myeloma)

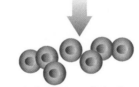

spleen and myeloma cells fuse to form hybridoma cells

grow hybridoma cells in tissue culture and select antibody-forming cells

collect monoclonal antibodies

One application of monoclonal antibodies is in the treatment of cancer. It is possible to attach a radioactive toxin to an antibody that can kill cancer cells.

Use this information, together with the diagram, to explain how monoclonal antibodies could be used to treat a patient suffering from lung cancer. *(6 marks)*

✓ ✓ ✓ ✓ ✓ ✓ Any six from:

extract cancer cells (from human lung tissue)

vaccinate mouse with cancer cell / antigen

collect spleen cells from mouse which contain **specific antibodies to the cancer / antigen**

join / fuse myeloma cells to spleen cells / lymphocytes that are sensitised / able to act upon cancer antigen

grow hybridoma cells and extract **specific antibodies which act on cancer** antigens

attach radioactive toxin to monoclonal antibodies;

inject patient with monoclonal antibodies

antibodies will attack **only the cancer cells** in the patient's body.

> Notice that in this answer it is not enough to just copy out the stages of monoclonal antibody production, you need to apply the knowledge to this particular situation – hence the sections above that are in bold-type. This type of question might appear as a 'level of response type' also, so make sure your spelling, punctuation and grammar – and in particular, the clarity of your explanations – is correct. It is worth drafting out this type of answer in rough on spare blank sheets in your exam paper before committing the answer to the space provided.

Plant Diseases

This section (beginning on page 52) includes details of tobacco mosaic virus as a viral disease, black spot as a fungal disease and aphids as insects. You will be expected to recall information about these examples.

The next question relates to an example that is not on the AQA specification but, as with many questions, all the necessary information is given. All you have to do is apply what you know about disease transmission and draw conclusions from data.

HT Example

Ash dieback, or Chalara, is caused by a fungus called *Hymenoscyphus fraxineus*. Chalara results in loss of leaves, crown dieback and bark damage in ash trees. Once a tree is infected, the disease is usually fatal because the tree is weakened and becomes prone to pests or pathogens.

a) Damage to leaves can be caused by lack of certain minerals that plants need.

Write down **one** mineral deficiency and how it can affect leaves. *(2 marks)*

✓ ✓ Accept any one deficiency and effect:

lack of nitrates causes yellow leaves and stunted growth

lack of magnesium causes chlorosis

lack of potassium / phosphate causes discolouration of the leaves.

The map gives an indication of where cases of Chalara were reported in 2012 in the UK.

Key:
■ = infection confirmed

Scientists have discovered that:

- Chalara spores are unlikely to survive for more than a few days
- spores can be dispersed by winds blowing from mainland Europe
- trees need a high dose of spores to become infected
- there is a low probability of dispersal on clothing or animals and birds.

b) Which one of the following conclusions is supported by evidence from the map?

Tick **one** box. *(1 mark)*

Chalara is limited to the East of England. ☐

Spores of Chalara arrived in England by being carried on winds from Europe. ☐

There is a high concentration of Chalara cases in the East of England. ☐

Ash trees in north-west Scotland are resistant to Chalara. ☐

There is a high concentration of Chalara cases in the East of England ✓.

c) One scientist suggests that cutting down and burning infected trees could eradicate the disease.

i) Explain how this method could be effective. *(2 marks)*

✓ ✓ Any two from:

burning trees destroys fungus / Chalara

prevents further spores being produced

reduces spread of spores.

ii) Give **one** reason why this control method may not stop the spread of Chalara. *(1 mark)*

✓ Any one from:

not all trees removed

trees may produce spores before being detected / destroyed

more spores could be introduced by winds from mainland Europe.

Plant Defence Responses

This section is largely about adaptations, so you will need to keep at the forefront of your mind the importance of linking features to advantages, as the example on the next page shows.

Example

The pictures below show two strategies used by plants to deter pests and herbivores.

Foxglove: contains a chemical that lowers heart rate.

Passion flower: possesses structures that look like butterfly eggs.

Explain how effective each of these plant's adaptations are in protecting them against being eaten by **different types of herbivore**. *(3 marks)*

✓ ✓ ✓ Any three of the below, with a maximum of 2 marks for one plant.

Foxglove

when eaten, the heart is affected, leading to heart attack or lack of oxygen to brain further eating by herbivore's mates or herd deterred as a result

only effective against vertebrates / animals with a heart / ineffective against invertebrates.

Passion flower

egg-like structures deter butterflies from laying eggs / butterflies do not want to lay eggs on a flower already with eggs / avoiding competition for the offspring

no caterpillars will hatch

only effective against insects / arthropods

*this is an example of **mimicry**.*

Notice that to gain full marks in this question you need to comment on **both** examples. A further clue to gaining extra marks is where the question asks you to address different types of herbivore.

For more on the topics covered in this chapter, see pages 34–41 of the *Collins GCSE AQA Biology Revision Guide*.

Photosynthesis

Photosynthesis is described as an endothermic reaction, meaning that it requires an energy input, in this case from sunlight. The energy is transferred to chloroplasts in the leaves. At the heart of understanding this topic is the photosynthesis equation, and hardly an exam goes by where this basic chemistry is not tested. So, it is well worth your while learning both word and symbol equations 'off by heart'.

Example

In the 17th century a Flemish scientist called Van Helmont carried out some experiments involving weighing the mass of a willow tree over five years. He found that the mass of the tree increased by 30 times and yet the soil mass remained constant.

a) If the tree had a starting mass of 120 g, calculate the finishing mass. Show your working. Show your answer in kilograms. *(2 marks)*

3.6 kg ✓ ✓

2 marks for correct answer, but if incorrect, 30 × 0.120 kg would gain 1 mark.

b) Van Helmont concluded that the willow's mass was entirely due to water intake. Explain why Van Helmont was only partially correct. *(1 mark)*

Carbon dioxide is also incorporated from the atmosphere ✓.

c) What, apart from water, might the willow have absorbed from the soil? *(1 mark)*

Mineral ions / nutrients ✓

d) Explain why this mass was hardly detectable. *(1 mark)*

Minerals are absorbed in small amounts ✓.

This is another example of a question incorporating aspects of scientific discovery and the scientific process. As you practise more of these you will see similar principles appearing time and again; such as a calculation, interpretation of observations and linking your biological knowledge to the experimental findings.

Bioenergetics

Example

Glucose can be used by plants for
energy or to build up bigger molecules.
The diagram shows a starch molecule.
The part labelled **A** is a glucose molecule.

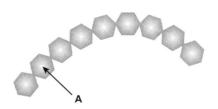

A

a) In which organs of the plant would most of this starch be manufactured? *(1 mark)*

   Leaves ✓

b) The plant can synthesise other molecules from the glucose it manufactures, such as
   cellulose and protein. State **one** use for cellulose and **one** use for protein. *(2 marks)*

   Cellulose: cell walls for support ✓.
   Protein: growth / cell membranes / enzymes ✓.

c) Write the word equation for photosynthesis. *(2 marks)*

   carbon dioxide + water ⟶ *glucose + oxygen* ✓ For reactants ✓ For products

The Rate of Photosynthesis

One of your required practicals is to investigate the effect of light intensity on the
rate of photosynthesis using an aquatic organism such as pondweed. The following
question is based on this practical.

Example

Some students were asked to carry
out a similar practical, only this
time they were asked to investigate
the effect of temperature on
photosynthesis rate in pondweed.
They set up the equipment shown
and changed the temperature
using ice and hot water. They
counted the number of bubbles
given off every minute at different
temperatures.

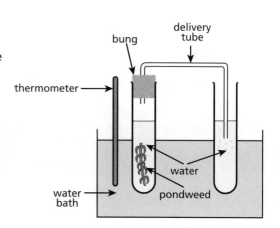

a) Why did the students use a water bath? *(1 mark)*

✓ Any one from:

to keep the water temperature constant

to vary the temperature for each trial.

b) What should the students have done to make the investigation fair? *(1 mark)*

Keep all variables (except the one being investigated) the same, e.g. same amount of pondweed for each experiment, same amount of water for each experiment, same light intensity ✓.

c) How could the students make sure their results were reproducible? *(1 mark)*

Repeat their investigation ✓.

d) Which gas was given off? *(1 mark)*

Oxygen ✓

e) HT The class had to pack up early and didn't finish gathering their data, but they were able to plot a graph (below) from the results they did obtain.

Continue the line on the graph to show the trend you would expect for temperatures above 30°C. *(1 mark)*

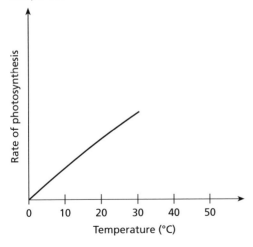

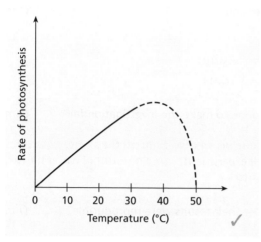

f) **HT** Explain why rate of photosynthesis varies with temperature over the range 0°C to 50°C. *(2 marks)*

As the temperature increases, the rate of photosynthesis increases due to more rapid molecular movement and therefore more successful collisions between molecules ✓. By 37–38°C the rate peaks and beyond this point the enzymes controlling photosynthesis become denatured and the reaction stops ✓.

g) **HT** A market gardener puts a wood-burning stove in his greenhouse to increase the yield of his lettuces.

i) Explain how this will increase the yield. *(2 marks)*

✓ ✓ Any two from:

the increased temperature from the stove will increase the photosynthesis rate

increase in carbon dioxide concentration will have the same effect

increased photosynthesis means increased starch / glucose / carbohydrate production / yield.

ii) Suggest **one** other measure he could take to increase his yield. *(1 mark)*

✓ Any one from:

increase light regime, e.g. artificial lighting switched on at night time

increase light intensity / brighter lights.

Higher Tier questions relating to rates of photosynthesis often touch on the idea of **limiting factors**. These factors are temperature, light intensity and carbon dioxide concentration. In general terms, when any one of these variables is increased, the rate of photosynthesis goes up. But this does not continue indefinitely. A maximum rate will be reached and, in the case of temperature, the rate will then fall off due to enzymes becoming denatured. Once the maximum rate is reached, another variable is acting as a limiting factor. In other words, if that factor was increased, the rate would continue to rise again.

Questions of this type are also linked to market gardening where each of these factors can be adjusted to maximise yields.

Respiration

Respiration is not the same as breathing (sometimes called 'ventilation'). Breathing delivers air with high concentrations of oxygen into the lungs where the oxygen is absorbed into the blood. Cells then use the oxygen and react it with glucose to release energy. Beware of describing energy as being created or made. It is better to think of it as changing from one form to another, i.e. from chemical energy to movement, heat or electrical (nervous conduction).

Example

a) Which substance is a product of anaerobic respiration in humans?
Tick **one** box. *(1 mark)*

Carbon dioxide	☐	Glucose	☐
Ethanol	☐	Lactic acid	☐

Lactic acid ✓

b) Which substance is a product of aerobic respiration in plants?
Tick **one** box. *(1 mark)*

Carbon dioxide	☐	Glucose	☐
Ethanol	☐	Lactic acid	☐

Carbon dioxide ✓

Bioenergetics

c) Niamh is training for a marathon. Every few days she runs a long distance. This builds up the number of mitochondria in her muscle cells.

What is the advantage for Niamh of having extra mitochondria in her muscle cells? Tick **one** box. *(1 mark)*

Her muscles become stronger. ☐

Her muscles can contract faster. ☐

Her muscles can release more energy. ☐

Her muscles can repair faster after injury. ☐

Her muscles can release more energy. ✓

d) Bob has been running hard and has an oxygen debt. Describe what causes an oxygen debt after a session of vigorous exercise and how Bob can recover from its effects. *(4 marks)*

Cause: build-up of lactic acid (in muscles) ✓ *due to anaerobic respiration* ✓.

Recovery: heavy breathing / panting ✓, *over a period of time* ✓.

e) Tariq is competing in a 10-mile running race. His heart rate and breathing rate increase. Describe how this helps his muscles during the race. *(3 marks)*

✓ ✓ ✓ *Any three from:*

more blood to the muscles

more oxygen supplied to muscles

increased energy transfer in muscles / avoids or reduces anaerobic respiration

more lactic acid is removed from muscles.

Metabolism

Metabolism refers to the sum of all chemical reactions in the body. This includes exothermic (or catabolic) reactions, which release energy, and endothermic (anabolic) reactions, which take energy in. Respiration is the main catabolic reaction in living systems.

HT Example

a) Isaac is running a marathon. Write a balanced symbol equation for the main type of respiration that will be occurring in his muscles. *(2 marks)*

$C_6H_{12}O_6 + 6O_2$ ✓ $\longrightarrow$ $6CO_2 + 6H_2O + energy\ released$ ✓

b) Isaac's metabolic rate is monitored as part of his training schedule. He is rigged up to a metabolic rate meter. This measures the volumes of gas that he breathes in and out.

The difference in these volumes represents oxygen consumption. This can be used in a calculation to show metabolic rate.

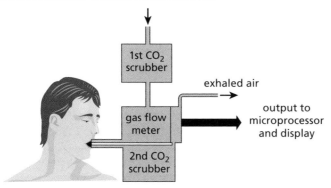

Here are measurements taken from the meter over a period of 1 hour.

	5 minutes of jogging	5 minutes of rest	5 minutes of sprinting	5 minutes sprinting on an incline
Mean metabolic rate per ml oxygen used per kg per min	35	20	45	60

i) The units of metabolic rate are expressed in the table as 'per kg'.

Why is this adjustment made? *(2 marks)*

✓ ✓ Any two from:

larger athletes will use more oxygen due to their higher muscle mass

the adjustment allows rates to be fairly / accurately compared

different athletes may have different masses.

ii) Using the table, explain the difference in readings for jogging and sprinting.

(2 marks)

Sprinting has a greater energy demand ✓, so more oxygen is needed ✓.

iii) Isaac does quite a lot of exercise. His friend Boris does not. How might Boris's readings compare with Isaac's? Give a reason for your answer. *(2 marks)*

Boris's metabolic rate would be lower ✓ because his lungs, heart and muscles are less efficient at transporting / using oxygen ✓.

In this Higher Tier example, the main thing to focus on is interpreting the information in the table. Key to this is understanding the units used for measuring metabolic rate. Oxygen consumption has been chosen because it is the most easily measurable of the reactants in respiration. You may know from chemistry that when assessing reaction rates you can either measure loss of reactants or gain in products.

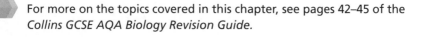

For more on the topics covered in this chapter, see pages 42–45 of the *Collins GCSE AQA Biology Revision Guide.*

Homeostasis and Response

Homeostasis and Negative Feedback

Control systems in the human body share a lot of comparisons with control systems in chemistry and physics, and you will come across these ideas if you ever study thermostats or biochemical systems. They share three components: receptors / sensors, coordination centres / comparators and **effectors**.

The terms 'homeostasis' and 'negative feedback' are often confused. Homeostasis simply refers to the way in which the body tries to keep certain factors constant, e.g. temperature, blood sugar levels. Negative feedback is a Higher Tier concept and is the mechanism by which the body's internal environment is achieved. The following questions explore these two ideas.

Example

Here is some information about how conditions are kept stable in the human body.

Write down the missing words. Choose **three** words from the box.   *(3 marks)*

effectors   spine   receptors   homeostasis   hormones   glands

Certain factors have to be kept constant in the body. This is achieved by a process

called _____. In order for this to happen, the central nervous system (CNS)

needs to receive information from the environment. This is accomplished through

_____ such as the eye or ear. Once the information has been relayed, the

CNS brings about appropriate changes through _____.

homeostasis ✓; receptors ✓; effectors ✓

Example

Match the structures below, with the function they perform.   *(3 marks)*

Pancreas	Skin receptor	Pituitary gland	Retina

releases ADH	detects pressure	detects light	produces insulin

Homeostasis and Response

All four correct = ✓ ✓ ✓, three correct = ✓ ✓, one or two correct = ✓.

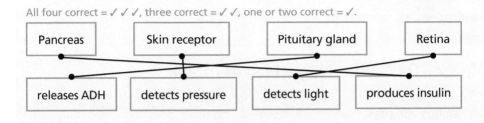

Pancreas	Skin receptor	Pituitary gland	Retina

releases ADH	detects pressure	detects light	produces insulin

Example

The diagram below shows how the hormone thyroxine is regulated in the body.

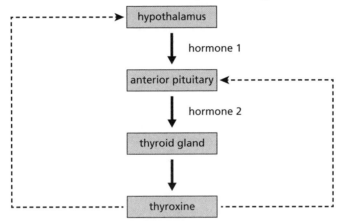

Thyroxine stimulates the basal metabolic rate. It plays an important role in growth and development.

a) What name is given to this process, where a system resists a change from a norm (set point) level? *(1 mark)*

> Negative feedback ✓

b) The **pituitary** is often referred to as the 'master gland' of the body because it stimulates other glands to release chemical messengers as well as producing many hormones of its own. Name **one** hormone released by the pituitary. *(1 mark)*

> ✓ Accept one from:
>
> LH
>
> FSH
>
> ADH
>
> These are hormones required by the specification but others might include ACTH, TSH.

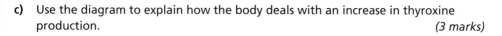

c) Use the diagram to explain how the body deals with an increase in thyroxine production. *(3 marks)*

✓ ✓ ✓ Any three from:

more thyroxine in blood detected by hypothalamus

less hormone 1 produced

less hormone 2 produced

negative feedback on hypothalamus and pituitary

fall in thyroxine level.

d) In a condition called hypothyroidism, not enough thyroxine is produced. Symptoms in adults include fatigue, low heart rate and weight gain. Use the information above to explain why these symptoms occur. *(3 marks)*

✓ ✓ ✓ Any three from:

lower metabolic rate lowers heart rate

therefore, less oxygen / nutrients reaching brain

fatigue caused by lack of oxygen / nutrients

lower metabolic rate means less breakdown of carbohydrate

more carbohydrate converted to fat.

Neurones and Reflex Arcs

Questions about this section of the specification often combine features of nerve cells together with how they operate in terms of the reflex arc. If you learn the general sequence below then you can apply it to any number of scenarios. Popular ones include the knee-jerk reflex, pupil reflex and the pain reflex.

Sense organ	Sensory neurone	Synapse	Relay neurone	Synapse	Motor neurone	Muscle
Receptors detect a change either inside or outside the body. This change is a stimulus.	Conducts the impulse from the sense organ towards the CNS.	The gap between the sensory and relay neurones.	Passes the impulse on to a motor neurone.	The gap between the relay neurone and the motor neurone.	Passes the impulse on to the muscle (or gland).	The muscle will respond by contracting, which results in a movement. Scientists call muscles (and glands) effectors.

Homeostasis and Response

Example

Merrick has been cooking and left a plate on a hot cooker ring. He picks the plate up and immediately drops it. His response is controlled by a reflex action.

a) What is the stimulus in this reflex? *(1 mark)*

Heat (from the hot plate) ✓

b) What is the effector? *(1 mark)*

(Arm) muscle ✓

c) Give **two** characteristics of a reflex action that makes them distinct from other types of response in the body. *(2 marks)*

Rapid ✓

Unconscious / do not require higher thinking or processing of information ✓.

d) Once Merrick picks up the plate, a nervous impulse is passed from his skin receptor along a sensory neurone. Describe the remainder of the pathway that the impulse travels to its completion at the effector. *(4 marks)*

✓ ✓ ✓ ✓ Any four from:

sensory neurone carries impulse to relay / intermediate neurone in CNS / spine

impulse passed on to motor neurone

motor neurone carries impulse to effector / muscle

reference to synapse / nerve junction.

Notice that this is a description, so you gain marks for simply outlining the route that the nervous impulse takes, using correct terminology. There is no need to explain **how** this happens. This specification does not require you to understand transmission of the impulse at the synapse level.

Reaction Times

One of your required practicals is to plan and carry out an investigation into the effect of a factor on human reaction times. Here is a question to help you see how you might need to draw on this information in an exam.

Example

The reaction times of six people were measured.

They put on headphones and were asked to push a button when they heard a sound. The button was connected to a timer.

Five trials of the experiment were carried out. The table below shows the results.

Person	Gender	Reaction time in seconds				
		1	2	3	4	5
A	Male	0.26	0.25	0.27	0.25	0.27
B	Female	0.25	0.25	0.26	0.22	0.24
C	Male	0.31	1.43	0.32	0.29	0.32
D	Female	0.22	0.23	0.25	0.22	0.23
E	Male	0.27	0.31	0.30	0.28	0.26
F	Female	0.23	0.19	0.21	0.21	0.22

a) Describe a pattern in these results. *(1 mark)*

Female reaction time better than male / females have faster reaction times than males ✓.

b) Calculate the mean reaction time for person C, ignoring any outliers. Show your working. *(2 marks)*

$\frac{0.31 + 0.32 + 0.29 + 0.32}{4}$ ✓

Answer is 0.31 ✓.

You are told to ignore any outliers. These values are so far outside the main trend of data as to be considered anomalous and most likely due to error. In this case the number 1.43 is vastly different to the other values.

c) When person D was concentrating on the test, someone touched her arm and she jumped. Her response was a reflex action. What are the **two** main features of a reflex action? *(2 marks)*

> Rapid / fast ✓
>
> Automatic / done without thinking ✓

The Brain

This section of the specification is relatively new and initially calls on you to learn the basic structure of the brain and the functions of the various parts.

Example

The diagram shows the human brain.

a) Write down the name of the part of the brain labelled X. *(1 mark)*

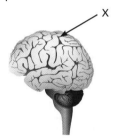

> X is the cerebral cortex / cerebrum ✓.

b) Which statements about the brain and its parts are correct? *(2 marks)*

Tick **two** boxes.

The cerebellum is responsible for controlling heartbeat. ☐

The brain contains junctions between all three types of neurone. ☐

The brain is part of the central nervous system. ☐

The medulla controls higher mental functions. ☐

> The brain contains junctions between all three types of neurone. ✓
>
> The brain is part of the central nervous system. ✓

Example

Neuroscientists have linked particular areas of the brain to different functions. Outline **one** technique used to map the brain and discuss an ethical issue associated with it. *(2 marks)*

✓ Any one technique from:

MRI scan

electrically stimulating the brain

study of people with brain damage

dissection of brain from dead person.

✓ Any one ethical issue from:

must not cause damage to patient's brain

patients may not be able to give permission.

The Eye

This is the main sense organ covered on the AQA specification and focuses on three main areas: the structure of the eye, focusing or accommodation, and eye defects. When describing functions of the eye, students often confuse and misunderstand cornea, lens, iris, pupil and retina. They also assume that the lens is the only structure involved in focusing. Here is a guide to take you through this information.

Cornea – transparent front part of the eye. It lets light through and refracts it (not reflects!).

Iris – a muscle that adjusts the size of the pupil (a hole in the eyeball). This is there to change the amount of light let into the eye. Too much light can damage the retina. Too little light will leave vision limited in dim light.

Lens – fine focuses the light. In accurate ray diagrams of the eye, the straight lines representing light rays should first bend at the cornea and then again at the lens. The lens changes shape by contraction of the circular muscle in the ciliary body / muscle. A fat or more convex lens is needed for objects that are close to the eye as light needs to be refracted more. The lens is stretched thin (less convex) for objects that are further away.

Retina – contains millions of light-sensitive cells. Light needs to be brought to a fine focus here (light rays meet exactly) in order for a clear image to be seen. The light energy is then transferred into electrical energy in the light-sensitive cells and this makes up the nervous signals that are sent down the optic nerve to the brain for interpretation.

Example

The eye is an example of a sense organ.

a) Match **A**, **B**, **C** and **D** with labels **1–4** on the diagram. Enter the numbers in the correct boxes. *(4 marks)*

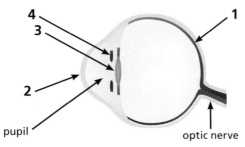

pupil

optic nerve

A lens ☐

B retina ☐

C cornea ☐

D iris ☐

A 3 ✓ B 1 ✓ C 2 ✓ D 4 ✓

b) This diagram shows the eye focusing on a distant object. Describe what happens to the light in order for it to be focused on the retina.

In your answer, name the parts of the eye involved. *(2 marks)*

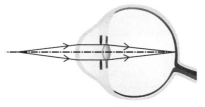

Light rays fall on the cornea where they are refracted / bent ✓; the lens then refracts / bends light further ✓.

c) Short sightedness or myopia is a defect of the eye.

Describe what causes myopia and explain how it can be corrected. *(2 marks)*

Caused by an eyeball that is too long / weak suspensory ligaments that cannot pull the lens into a thin shape ✓; corrected using a concave lens ✓.

The table shows some data obtained from a study of retinas in different cat species. Cones are cells on the retina that enable animals to see in colour. Rods are not as sensitive to different colours but are more sensitive to light intensity.

Species	Cone density / per mm²
Bobcat	100 000
Domestic cat	40 000

From looking at the table, Emelia concluded that bobcats have better eyesight than domestic cats.

d) Do you agree? Give reasons for your decision, including its limitations. *(3 marks)*

Bobcats are likely to be able to form more detailed images in bright light / have better colour vision ✓ because cone density is greater ✓. Cannot conclude that they have better night vision as data about rods is not available ✓.

Control of Body Temperature

Control of body temperature is achieved through the thermoregulatory centre in the brain. It contains receptors that detect temperature changes in the blood. The skin contains temperature receptors that feed in nervous information to the thermoregulatory centre. The effectors are many but, as usual, consist of various muscles such as circular muscles in arterioles (during vasoconstriction), or erector pili muscles in the skin, which raise hairs.

The principles of control discussed at the start of this section can be seen in the following example.

Example

Rafiq is enjoying a skiing holiday, but not the cold! His body is working hard to keep at a constant temperature.

a) After standing while waiting for a ski lift, Rafiq starts to shiver. Explain how shivering helps him to maintain his temperature. *(2 marks)*

✓ ✓ Any two from:
muscle contraction
generates heat
through respiration.

b) After skiing cross-country for a while, Rafiq starts to sweat underneath his thermals. Explain how sweating enables him to lose heat. *(2 marks)*

> ✓ ✓ Any two from:
> evaporation of sweat from skin
> requires heat from body
> radiation
> endothermic change.

c) Rafiq's internal body temperature has changed very little during the day. What temperature is this likely to be? Circle the correct answer. *(1 mark)*

| 30°C | 35°C | 37°C | 40°C | 50°C |

> 37°C ✓

d) Rafiq's blood vessels can undergo vasoconstriction. Explain how this helps him conserve heat. *(3 marks)*

> ✓ ✓ ✓ Any three from:
> blood vessels / arterioles narrow
> blood shunted below adipose tissue / at layer
> fat insulation reduces heat loss from blood
> less blood flows close to the skin.

> Be careful when explaining loss of heat through sweating. It is often mistakenly suggested that the water in sweat is cool and therefore transfers heat away from the skin simply by virtue of this. It is important to explain the process of evaporation as being endothermic, as stated in the answer. You may have carried out cooling experiments in the laboratory where alcohol is rubbed over the surface of a flask containing hot water and a thermometer. You can observe first-hand how quickly the evaporating alcohol absorbs the heat. If you happen to get any of the alcohol on your skin, you can physically feel it getting cooler!

The Endocrine System

As well as the nervous system, the endocrine system makes up the other method of control available to the human body. Key points include:

- information / signalling is achieved through chemical messengers called hormones rather than nervous impulses

- response times are slower as these hormones have to diffuse to target organs through the bloodstream, which requires time

- hormones often have a more general effect on many cells, tissues and organs

- the response is more drawn out.

Example

Complete the missing information in the table, which is about different endocrine glands in the body. *(4 marks)*

Gland	Hormones produced
Pituitary gland	and
Pancreas	Insulin and glucagon
	Thyroxine
	Adrenaline
Ovary	and
Testes	Testosterone

1 mark for each correct line.

Gland	Hormones produced
Pituitary gland	**TSH, ADH, LH** and **FSH** ✓ Accept any two.
Pancreas	Insulin and glucagon
Thyroid gland ✓	Thyroxine
Adrenal gland ✓	Adrenaline
Ovary	**Oestrogen** and **progesterone** ✓
Testes	Testosterone

Control of Blood Sugar and Diabetes

This area of control illustrates how homeostasis and negative feedback operate in a specific example. Higher Tier knowledge is required to explain the role of the additional hormone, glucagon, in negative feedback.

Example

A new nanotechnology device has been developed for people with diabetes – it can detect levels of glucose in the blood and communicate this information to a hormone implant elsewhere in the body. The implant releases a precise quantity of hormone into the bloodstream when required.

a) Explain how this device could help a person with type 1 diabetes who has just eaten a meal. *(2 marks)*

> ✓ ✓ Any two from:
>
> *after a meal, a rise in glucose levels will be detected by device*
>
> *which will cause hormone implant to release insulin*
>
> *insulin released to bring blood glucose level down.*

b) Explain why a person with type 2 diabetes might not have as much use for this technology. *(2 marks)*

> *People with type 2 diabetes can often control their sugar level by adjusting their diet ✓; body's cells often no longer respond to insulin ✓.*

In questions about diabetes, make sure you refer to blood glucose as glucose is found all over the body, but it is in blood that the control system operates.

HT Example

The word equation below shows how two hormones produced in the pancreas regulate blood glucose in the body.

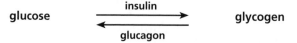

glucose ⟶ insulin ⟶ glycogen
glucose ⟵ glucagon ⟵ glycogen

Lazlo has just eaten a doughnut before spending an hour playing a game of basketball.

Use the equation to describe and explain how negative feedback operates to ensure that his blood glucose levels remain within acceptable limits over this period of time. *(4 marks)*

✓ ✓ Any two from:

glucose levels in blood raised after doughnut

changes detected (in the pancreas)

insulin released from cells in pancreas

stimulate body cells to absorb glucose

also stimulate liver and muscle cells to convert glucose to glycogen.

✓ ✓ Any two from:

blood glucose levels fall during the basketball game

changes detected in pancreas (mark available if not already gained above)

glucagon released from different cells in pancreas

this hormone stimulates liver and muscles to convert glycogen to glucose.

It is important not to confuse hormones with enzymes. Although they are proteins they have a different mode of action. The three 'g' words can also be confusing: glucose – the sugar, glycogen – the storage carbohydrate, and glucagon – the hormone. Be aware that some questions about blood glucose control may ask you to interpret graphs showing these changes in someone's blood over time. Comparing diabetics with non-diabetics is also a possibility.

Water and Nitrogen Balance

Control of water and nitrogenous compounds is exercised via the kidneys. Although you won't be asked directly about the structure of the kidneys and nephrons / kidney tubules it helps to be aware of these as it aids in your understanding of these difficult ideas.

 Example

The element nitrogen takes many forms in the body including the molecules: amino acids, ammonia and urea. When protein is broken down in the digestive system it produces amino acids, which are essential for the body in producing its own specific proteins. However, there are often excess amino acids that cannot be stored. Explain what happens during the process of **deamination,** stating why it is important that any products are disposed of. *(3 marks)*

✓ ✓ ✓ Any three from:

amino acids are deaminated in the liver

ammonia produced

which is toxic

converted to urea (for excretion).

Homeostasis and Response

Example

A marathon runner is resting the day before she competes in a race. The table shows the water that she gains and loses during the day.

Method gained	Amount of water gained (ml)	Method lost	Volume of water lost (ml)
In food	1000	In urine	
From respiration	300	In sweat	800
Drinking	1200	In faeces	100
Total gained		**Total lost**	2500

a) How much water does the runner lose in urine during the day? *(1 mark)*

   1600 ml ✓

b) What can you say about the total water gained and the total water lost in the day? Why is this important? *(2 marks)*

   Amounts are equal ✓; important that water intake should balance water output to avoid dehydration ✓.

c) The runner runs a marathon the next day. Suggest and explain how the figures shown in the table may alter during the day of the race. *(4 marks)*

   Intake of water greater ✓; output from sweating greater ✓; water gained from respiration greater ✓, as muscles contracting more / respiring more ✓.

Kidney disease often comes up as a subject for questions. There are two methods currently available for dealing with malfunctioning kidneys: dialysis and kidney transplants. You should be aware of the advantages and limitations of each.

Example

a) How might the composition of urine in a healthy person differ from that of a person suffering from kidney disease? *(2 marks)*

   Water content falls ✓; ion balance cannot be regulated – any reference to sodium, chloride, potassium, etc. urea concentrations rise / proteins present in urine ✓.

b) A person with kidney disease can be treated through dialysis or by receiving a transplanted kidney. Compare the advantages and disadvantages of these two methods. *(4 marks)*

✓ Any one from:

Dialysis advantages

does not require donor

shorter waiting list

does not require surgery

no need for immune-suppressant drugs.

✓ Any one from:

Dialysis disadvantages

must be carried out at least three times a week / regularly

temporary solution

expensive for NHS long-term

restrictions on diet between dialysis sessions.

✓ Any one from:

Transplant advantages

person can lead a more or less normal life without restrictions on diet and activity

less expensive for NHS. (If marking point not gained above.)

✓ Any one from:

Transplant disadvantages

must take immuno-suppressant drugs / increased risk of infection (If marking point not gained above.)

shortage of organ donors

kidney only lasts 8–9 years

operation carries risks.

The interplay of organs and hormones in the body's regulation of water levels is complex, and it is worth exploring common areas of misunderstanding. Take time to study diagrams that summarise this control system and practise explaining scenarios where the water potential of blood falls and rises.

Homeostasis and Response

HT Example

The body's water levels are monitored by assessing the osmotic or water potential of the blood.

Belinda has just eaten a meal of fish and chips containing salt and vinegar. Some time later she follows this by drinking a pint of water. Explain how the body's control system ensures that the water potential of Belinda's blood remains within acceptable levels.

In your answer, outline any organs involved and suggest how the concentration of her urine would change.   (6 marks)

This is a 'level of response' question and, therefore, a typical mark scheme is not employed. Below is an exemplar that would gain full marks. The key points are:

- monitoring of water potential in blood by brain – specifically the hypothalamus
- rise in water potential of blood / dilution results in more ADH being released by the pituitary
- fall in water potential of blood results in less production of ADH
- ADH increases the permeability of the collecting duct in the kidney
- more ADH production means more concentrated urine.

✓ ✓ ✓ ✓ ✓ ✓ This is a model answer that would score the full 6 marks.

When Belinda's meal of fish and chips is digested, the salt is absorbed into her bloodstream and lowers the water potential of the blood. This fall is detected by the hypothalamus / brain, which sends impulses to the pituitary gland. The pituitary then secretes more ADH into the blood which, in turn, carries the chemical messenger to the kidney. Here, ADH causes the collecting duct to become more permeable and more water is reabsorbed into the bloodstream. This makes the urine more concentrated and raises the water potential of the blood. Upon drinking the water, the water potential of Belinda's blood rises. This sets in motion a reversal of all the above changes: less ADH produced by the pituitary, collecting duct less permeable to water, more water in urine therefore making it more dilute, water potential of the blood falls. This whole process is an example of negative feedback, allowing water levels in the blood to fluctuate by only small amounts around a norm level.

This answer covers more than enough marking points and it would be unlikely that you would have to specifically state all of the reverse details in the second part. However, they are all shown here for the sake of completeness. Please note that comparative terms are used throughout, i.e. more ADH not no ADH. This is a common theme in GCSE science questions.

Reproductive Hormones

Reproductive hormones are another complex area as the interplay of female hormones involves four chemical messengers: FSH, oestrogen, progesterone, and LH.

Example

The graph shows the thickness of the uterus during the menstrual cycle. Use the graph and your scientific knowledge to describe what happens in the woman's ovaries and uterus between days 5 and 28. *(3 marks)*

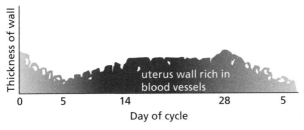

Days 5–14: uterus wall is being repaired ✓, egg released at approximately 14 days from ovary ✓, days 14–28: uterus lining maintained ✓.

HT Example

The diagram shows how the human process of ovulation is controlled.

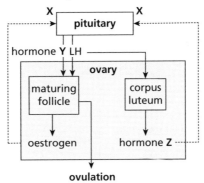

a) X represents an effect that two hormones have on the pituitary gland. Write down the name of this effect. *(1 mark)*

  Negative feedback ✓

b) Name hormone Y. *(1 mark)*

  FSH ✓

c) Name hormone Z. *(1 mark)*

  Progesterone ✓

d) Describe the effect on hormone Z if the egg is fertilised. *(2 marks)*

Continues to be produced ✓ in large quantities / at high levels ✓.

The best way to become confident in explaining changes in these processes is to use a generalised diagram and imagine various scenarios that occur during the woman's monthly cycle. Try following the arrows on the hormone flow diagram above and say aloud what would happen when one hormone level is raised. Repetition of this exercise will embed the information.

Contraception and Infertility Treatment

Tackling questions on contraception and infertility means that you will need a thorough understanding of the functions of female (and sometimes male) hormones. You will often find ethical considerations as part-questions.

Example

Tim and Margaret are finding it hard to conceive a child. They visit a fertility clinic and meet some other couples. The table shows some information about the problem that each couple has.

Couple	Problem causing infertility	Percentage of infertile couples with this problem	Percentage success rate of treatment
Tim and Margaret	Blocked fallopian tubes	13	20
Rohit and Saleema	Irregular ovulation	16	75
Leroy and Jane	No ovulation	7	95
Gary and Charlotte	Low sperm production	15	10
Ian and Kaye	No sperm production	21	10
Stuart and Mai	Unknown cause	28	–

a) Which couple has the best chance of being successfully treated? *(1 mark)*

Leroy and Jane ✓

b) In how many of the six couples is the problem known to be with the female? *(1 mark)*

> *Three: Tim and Margaret, Rohit and Saleema, and Leroy and Jane ✓.*

c) The treatment of irregular ovulation and no ovulation have the highest success rates.

Explain why treating irregular ovulation would produce more pregnancies in the whole population. *(2 marks)*

> *Although irregular ovulation has a lower success rate ✓, it affects over twice as many couples (16 × 75 produces a larger total than 7 × 95) ✓.*

d) Leroy and Jane are considering two methods to help them have children. The first is to have an egg donated by another woman. The second is to arrange for another woman to conceive the child using sperm from Leroy, then give birth to it (surrogacy). What are the advantages and disadvantages of each method? *(4 marks)*

> *Both methods mean that Jane does not make any genetic contribution ✓.*
> *Egg donation has a high rate of success but can be expensive ✓.*
> *Surrogacy might be cheaper but there is a risk that the surrogate mother might develop an attachment to the baby / want to keep it ✓.*
> *Egg donation requires invasive technique ✓.*

Example

Explain how the contraceptive pill works. Name any hormones involved. *(2 marks)*

> ✓ ✓ Any two from:
> *the contraceptive pill contains hormones that inhibit FSH production*
> *e.g. oestrogen / progesterone*
> *eggs therefore fail to mature*
> *progesterone causes production of sticky cervical mucus that hinders movement of sperm.*

Plant Hormones and their Uses

Questions usually cover three areas: tropisms and how plant hormones govern them, experimental evidence proving the mechanism by which tropisms operate, and applications of plant hormones for commercial purposes. There are many plant hormones but the one you need to be most aware of is **auxin**.

Example

Jan placed a bean seedling between a piece of damp filter paper and the wall of a glass jar. She observed it over a period of five days under constant illumination from uni-directional light. After this time, she drew a diagram of what she saw. Her observations are shown below. This is a plan view of the seedling.

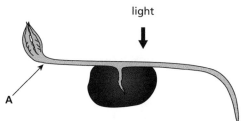

light

A

a) Which statements about the seedling are correct? Tick **two** boxes. *(2 marks)*

The root is showing negative phototropism. ☐

The shoot is showing positive geotropism / gravitropism. ☐

The seedling is not responding to light. ☐

Auxin has the same effect on the roots and shoots. ☐

Auxin is in greater concentration on the lower side of the root. ☐

Auxin has not been produced at all. ☐

The root is showing negative phototropism. ✓

Auxin is in greater concentration on the lower side of the root. ✓

Subtract 1 mark for every extra tick given to a maximum of –2.

b) Explain what has happened to cause the growth in the shoot at point **A** in the diagram. *(2 marks)*

Light causes auxin to build up on the lower side of the shoot / side of shoot away from light ✓, *which causes greater* **cell elongation** *on this side of the shoot* ✓.

Remember that light has a different effect on root and shoot tips. In roots, auxin (the hormone) causes less cell elongation whereas in shoots it causes more cell elongation. Do not refer to more cell division. The plant cells do not respond by multiplying.

For more on the topics covered in this chapter, see pages 46–55 of the *Collins GCSE AQA Biology Revision Guide*.

6 Inheritance, Variation and Evolution

Sexual and Asexual Reproduction

Reproduction is one of the seven characteristics of life, and at GCSE you will have studied well beyond the subject of human reproduction and considered the range of reproductive strategies that other organisms adopt. Remember, reproduction drives survival, which in turn drives evolution. So it is no wonder that these ideas are linked together in one topic. Much of the focus is on comparing sexual and asexual reproduction, and you need to learn the advantages and disadvantages of each in a thorough manner, then practise applying this knowledge to many examples.

Example

From the box below, choose **three** words to complete the sentences.   *(3 marks)*

zygotes	gametes	diploid	haploid	mitosis	meiosis

Eggs and sperm are known as sex cells or _____. They are described

as _____ because they contain one set of chromosomes. Eggs and

sperm are produced in the ovaries and testes by _____.

> Eggs and sperm are known as sex cells or **gametes**. They are described as **haploid** because they contain one set of chromosomes. Eggs and sperm are produced in the ovaries and testes by **meiosis**.

Example

Which statements about causes of variation are true? Tick **two** boxes.   *(2 marks)*

Meiosis shuffles genes, which makes each gamete unique. ☐

Gametes fuse randomly. ☐

Zygotes fuse randomly. ☐

Mitosis shuffles genes, which makes each gamete the same. ☐

> Meiosis shuffles genes, which makes each gamete unique ✓.
> Gametes fuse randomly ✓.

Inheritance, Variation and Evolution

Example

Sexual reproduction is the best strategy for organisms because it allows variation and therefore greater adaptation.

Asexual reproduction is better because when an organism is well adapted, it can produce exact copies of itself.

a) James and Ayesha disagree about which type of reproduction is most beneficial to organisms. Explain why both James and Ayesha's statements can be correct depending on the environmental conditions. Give the reasons for your choice. *(3 marks)*

✓ ✓ ✓ Any three from:

sexual reproduction can be an advantage to a species if the environment changes

asexual reproduction is more advantageous when the environment is not changing

some organisms use both types of reproduction, therefore both have their advantages

sexual reproduction might yield disadvantageous adaptations in an individual when the environment changes.

b) Apart from variation, write down **one** other difference between sexual and asexual reproduction. *(1 mark)*

✓ Any one from:

Sexual	Asexual
involves fusion of gametes	does not involve fusion of gametes
male and female parents required	only one parent required
slower than asexual	faster than sexual
requires meiosis	requires mitosis only
more resources (e.g. time, energy) required.	fewer resources required.

To gain the mark for part **b)** you must ensure that you have compared each type of reproduction with reference to the other, e.g. in sexual reproduction both male and female parents are required whereas in asexual reproduction only one parent is needed.

c) Describe how yeast carries out asexual reproduction. *(2 marks)*

Cytoplasm and organelles duplicated ✓, as a 'bud' ✓.

> It is impossible to separate sexual reproduction from variation and meiosis (covered in topic 1) so be prepared to field questions that come from these areas of the specification when dealing with reproduction questions.

More Reproductive Strategies

There are a few examples given by the specification that illustrate how asexual reproduction can occur in animals, plants and fungi. Also, the example of the malarial parasite crops up again (see page 54) because its life cycle is particularly complex.

Example

The diagrams below show two examples of asexual reproduction in plants. Firstly in the strawberry plant, and secondly in daffodil bulbs.

Strawberry plant

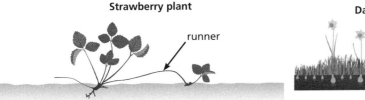

runner

Daffodils

bulb

a) What type of cell division is occurring in the strawberry runners and the daffodil bulbs to enable this type of reproduction? *(1 mark)*

Mitosis ✓

b) The daffodil produces reproductive structures called bulbs, which store food underground, whilst the strawberry plant sends out 'runners'. Suggest an advantage for each of these plant's strategies. *(2 marks)*

Food storage below ground helps the daffodil through the winter when conditions are too harsh for the flower above ground ✓.

The strawberry's runners allow it to spread rapidly so it can then produce flowers and reproduce sexually ✓.

c) Suggest **one** other advantage to these plants for reproducing in this way – compared to reproducing by sexual means. *(1 mark)*

> ✓ Any one from:
>
> *only one parent required / other parent plant to cross-pollinate with rapid reproduction and colonisation of new areas.*

d) The malarial parasite combines both sexual and asexual means of reproduction. Here is a summary diagram of its complex life cycle:

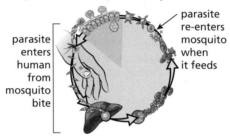

parasite enters human from mosquito bite

parasite re-enters mosquito when it feeds

i) The malarial parasite can live in two different organisms. Suggest a survival advantage in this. *(1 mark)*

> *Not as dependent on one species / if one species in low numbers or becomes extinct there is another organism available to act as a source of nutrients and 'shelter' ✓.*

ii) To control the spread of malaria, efforts have focused on attacking different parts of this life cycle. Explain how **one** method of control can be used to disrupt **one** part of the parasite's reproductive cycle. *(2 marks)*

> Focus: mosquito vector ✓
> ✓ Any one from:
> *kill adults with insecticide*
> *drain swamps where eggs are laid*
> *spread oil on stagnant water to prevent mosquito larvae living there*
> *use mosquito nets.*
> Or
> Focus: human host ✓
> ✓ Any one from:
> *vaccinate population against malaria*
> *treat patients with drugs that kill malarial parasite, e.g. quinine.*

DNA and the Human Genome Project

Foundation Tier knowledge of DNA covers basic DNA structure, the nature of DNA as a code in terms of base sequences, historical aspects of DNA's discovery, the Human Genome Project and the Genographic Project. If you learn the foundation ideas about DNA and what its function is, you should be able to apply these to questions that ask you to use your knowledge in novel situations. Remember that the bases A, C, T and G represent molecules in their own right and their sequence codes for the order of **amino acids**, which in turn make up **proteins** – the basis for the whole organism's body. Actual protein synthesis is a Higher Tier concept, so you won't need to learn this biochemistry unless you are sitting the Higher Tier papers.

Example

Which statements about the Human Genome Project (HGP) are true?
Tick **three** boxes.   *(3 marks)*

The genome of an organism is the entire genetic material present in its adult body cells. ☐

The data produced from the HGP produced a listing of amino acid sequences. ☐

The HGP involved collaboration between US and UK geneticists. ☐

The project allowed genetic abnormalities to be tracked between generations. ☐

The project was controversial as it relied on embryonic stem cells. ☐

The genome of an organism is the entire genetic material present in its adult body cells. ✓

The HGP involved collaboration between US and UK geneticists. ✓

The project allowed genetic abnormalities to be tracked between generations. ✓

Example

Studies of genomes can help scientists work out the evolutionary history of organisms by comparing the similarity of particular DNA sequences that code for a specific protein.

The table shows the percentage DNA coding similarity for protein A in different species.

Species	% DNA coding similarity between species and humans for protein A
Human	100
Chimpanzee	100
Horse	89
Fish	79
Yeast	67
Protist	57

a) What evidence is there in the table that closely related organisms developed from a common ancestor? *(1 mark)*

Organisms with very similar features / chimpanzees and humans share equal DNA coding for protein A ✓.

b) Using only the information from the table, which invertebrate is the most closely related to humans? *(1 mark)*

Yeast ✓

Example

The Human Genome Project has enabled specific genes to be identified that increase the risk of developing cancer in later life. Two of these genes are the *BRCA1* and *BRCA2* mutations.

a) If women are prepared to take a genetic test, how could this information help doctors advise women about breast cancer? *(2 marks)*

✓ ✓ Any two from:

warn women about the risk of cancer ahead of time

enable early and regular screening

enable early treatment

suggest treatment that is targeted.

b) If a woman possesses these mutations, it does not mean that she will definitely develop breast cancer. Why is this? *(2 marks)*

> *Other factors may contribute to onset of cancer* ✓*; risk is in terms of a probability (which is not 100%)* ✓*.*

Example

a) The picture below shows Max, Sharon and their three children: Dylan, Emily and Georgina.

Which members of the family share the least genes?
Tick **one** box. *(1 mark)*

Emily and Sharon	☐
Max and Dylan	☐
Max and Sharon	☐
Georgina and Max	☐

> *Max and Sharon* ✓

> Max and Sharon are unrelated genetically, whereas siblings share 50% of their DNA with each other and with each individual parent.

b) HT DNA contains four bases: G, C, A and T.

There are the same number of G bases as C bases. There are also the same number of A bases as T bases.

The analysis of one strand of DNA found that 27% of the bases were T. Calculate the percentage of the other three bases. Show your working. *(3 marks)*

A – 27% (as it is the same percentage as T) ✓

C – 23% ✓

G – 23 % ✓

To calculate C and G, we simply add together the percentages for A and T, which comes to 54%. This means that C and G together must account for 46% of the total DNA. Since C and G have the same percentage, we halve 46% to get 23%.

The bases that form the 'rungs' on the DNA ladder and code for the structure of a protein are arranged in pairs.

The figure below shows a diagram of DNA. One strand has been filled in.

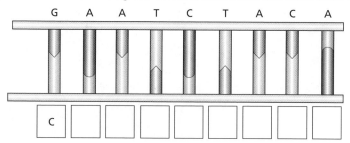

c) Enter the matching compounds in the boxes. *(1 mark)*

T, T, A, G, A, T, G, T ✓

d) How many amino acids are coded by the section of DNA shown in the figure? Explain your answer. *(3 marks)*

3 amino acids ✓

3 pairs of bases code for 1 amino acid ✓

9 base pairs = 3 sets of base pairs ✓

DNA and Protein Synthesis

Protein synthesis is Higher Tier material and it is often useful to visualise what is happening at a sub-cellular level to grasp what can initially be a difficult idea to understand.

It is useful to think of DNA as the book that carries the instructions for the functioning of, not only the cell, but the whole body. To express or realise the production of proteins that will carry out these functions requires a message to be sent to the production unit where the proteins are made – the **ribosome**.

The messenger is mRNA, which can be likened to a photocopy being taken of the relevant section of the book in the library (not the whole book!). The photocopy is interpreted by the ribosome and amino acids linked together according to the base sequence. Of course, there might be faults in the original code (library book) or even in the production of the photocopy. This is where **mutation** can cause disruption of protein production. These ideas are explored in the question below.

HT Example

Mutations occur when genes on DNA change so that they code for different proteins (or sequences of amino acids).

a)   State **two** environmental sources of mutation.   *(2 marks)*

> ✓ ✓ Any two from:
> UV light
> radioactive substances
> X-rays
> certain chemicals / mutagens.

b)   A change occurs in a section of DNA that leads to a new protein being formed.

Explain how this is possible and why the protein may not be able to perform its function.   *(3 marks)*

> Base / codon / triplet sequence changed ✓.
> Leads to change in amino acid sequence ✓; protein no longer has correct shape to perform its job ✓.

HT Example

The diagram on the next page shows the processes that occur in the cell when a DNA code is used to produce a protein. For each of the stages (1–4), describe and explain how the various molecules are involved.   *(4 marks)*

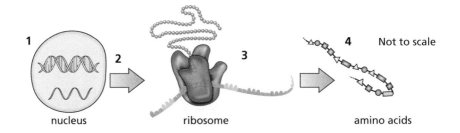

Stage 1: DNA unzips and exposes the bases on each strand / a molecule of messenger RNA (mRNA) is constructed from a template strand ✓.

Stage 2: The mRNA carries a complementary version of the gene with U (uracil) in place of T / travels out of the nucleus to the cytoplasm ✓.

Stage 3: In the ribosome, the mRNA is 'read' / tRNA molecules carry individual amino acids to add to a growing protein (polypeptide) ✓.

Stage 4: The new polypeptide folds into a unique shape / is released into the cytoplasm ✓.

Example

Some sections of DNA are **non-coding**. What is the purpose of these base sequences? *(1 mark)*

To switch genes on and off / determine how genes are expressed ✓.

Inheritance: Working out Genetic Crosses

Working out genetic crosses is a little like calculating maths problems. You understand the ideas behind them, grasp the constructs (arrow diagrams / Punnett squares) and then carry out many examples to get the hang of them. Before attempting these problems, make sure you understand the meaning of the following terms: **phenotype**, **genotype**, **recessive**, **dominant**, **heterozygous** and **homozygous**.

Generally, three types of 'cross' appear in questions. These could be called the 'pure breeding' cross, which yields only the dominant phenotype, the 'heterozygous' cross, which gives a 3:1 ratio of phenotypes, and the 'back-cross', which produces a 1:1 ratio of phenotypes. We will explore all three types of cross in the following examples.

Example

Raj is the owner of two dogs and both are about two years old. Both dogs are black in colour and came from the same litter of puppies.

a) A dog's adult body cell contains 78 chromosomes. How many chromosomes would be in a male dog's sperm cells? *(1 mark)*

39 ✓

b) The dogs' mother had white fur and the father had black fur. Using what you know about dominant alleles, suggest why there were no white puppies in the litter. *(2 marks)*

Black is the dominant gene / allele; white is recessive ✓. No marks given for references to 'black chromosome' or 'white chromosome'.

The allele for black fur is passed on / inherited from the father ✓.

c) HT One year later, one of the black puppies mated with a white-haired dog. She had four puppies. Two had black fur and two had white fur.

The letters **B** and **b** represent the alleles for fur colour: **B** for black fur and **b** for white fur.

Draw a fully labelled genetic diagram to explain this. Show which offspring would be black and which would be white. *(3 marks)*

	b	b
B	Bb Black	Bb Black
b	bb White	bb White

Or

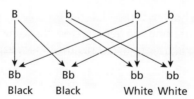

Correct genotype or gametes for both parents (Bb and bb) ✓
Genotype of offspring correct (Bb and bb) ✓
Correct phenotype of offspring ✓

Part **c)** of this question is an example of a 'back-cross.' For Foundation Tier, you will be provided with the Punnett square for working out a genetic cross. For Higher Tier, you are expected to construct your own. Remember to represent the alleles as single letters and to make the distinction between capital (dominant) and lower case (recessive) letters. Some letters cause particular difficulty, e.g. N and n.

Inheritance, Variation and Evolution

Lastly in this section, let's look at the heterozygous cross. A popular example is the inheritance of eye colour.

 Example

Explain how parents who both have brown eyes could produce a child who has blue eyes. Use a genetic diagram to help you, and state the probability of a brown-eyed child occurring. *(4 marks)*

If both parents are heterozygous for this gene, e.g. Bb ✓, then there is a 1 in 4 chance of the child having blue eyes ✓.

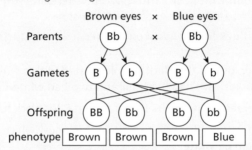

✓ ✓ for correct diagram. Subtract 1 mark for every incorrect row.

The above answer makes use of an 'arrows' diagram. You could equally have used a Punnett square.

Genetic Defects

This section will deal with how to tackle questions involving genetic abnormalities or defects. The principles of using the genetic cross diagrams still apply but the only difference is recognising and interpreting how the defect expresses itself.

The following example highlights the popular examples, **cystic fibrosis** and **polydactyly**, as these are required by the specification.

two

Example

Fill in the missing words to complete the following sentences.

Choose from the words below. *(4 marks)*

heterozygous	dominant	carriers	two
recessive	one	three	homozygous

Polydactyly is a condition that causes extra fingers or toes. It's caused by a

_____ allele. Only _____ parent needs to have the

disorder for a child to be affected.

> Polydactyly is a condition that causes extra fingers or toes. It's caused by a
> **dominant** allele. Only **one** parent needs to have the disorder for a child to
> be affected. ✓ ✓

Cystic fibrosis is caused by a _____ allele. It must be inherited

from both parents. The parents might not have the disorder, but they might be

_____ .

> Cystic fibrosis is caused by a **recessive** allele. It must be inherited from both
> parents. The parents might not have the disorder, but they might be **carriers**. ✓ ✓

A man who knows he is a carrier of cystic fibrosis is thinking of having children with
his partner. His partner does not suffer from cystic fibrosis, neither is she a carrier.
Suggest what a genetic counsellor might advise. Use genetic diagrams to aid your
explanation and state any probabilities of offspring produced. *(6 marks)*

✓ ✓ ✓ ✓ ✓ Any six from:

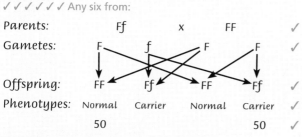

Parents: Ff   x   FF   ✓

Gametes: F   f   F   F   ✓

Offspring: FF   Ff   FF   Ff   ✓

Phenotypes: Normal Carrier   Normal Carrier ✓

              50   50   ✓

No children will suffer from cystic fibrosis ✓.

50% chance of a child being a carrier ✓.

Sex Determination

This area of genetics should be straightforward to understand, yet many students lose marks due to incorrectly stating that the male genotype is simply 'Y', whereas it is 'XY'.

Example

What is produced from the fusion of two sex cells? *(1 mark)*

A zygote ✓

Example

Which of the following are the female sex chromosomes, and which are the male sex chromosomes? Label them correctly. *(1 mark)*

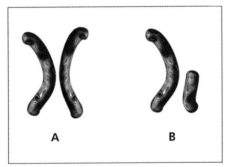

A   B

A: female, B: male ✓

Example

John and Salmyra decide to have children. Use a genetic cross diagram to show that the probability of them producing a girl is 50%. *(4 marks)*

Parents: John XY   x   Salmyra XX   ✓
Gametes: X   Y   x   X   X   ✓
Offspring:

	X	Y
X	XX	XY
X	XX	XY

✓

Ratio of male:female   1:1 ✓
(Probability of female child therefore 50%.)

Variation and Evolution

Variation, as far as GCSE biology is concerned, refers to differences in individuals within the same species. From KS3 work you will know that this can either be caused by the genes organisms possess, the influence of their environment, or both. Variation is closely linked to evolution because it drives the process called **natural selection** (the mechanism by which evolution occurs). This next example explores these connections.

Example

Scientists frequently study the distribution of the common snail, *Cepaea*. The snail has a shell that can be brown or yellow, and striped or unstriped. The shell colour and banding influences the visibility of snails to thrushes that prey on them.

A recent study compared the distribution of snails in forest and countryside areas. The results, as a percentage, are shown below.

Area	Striped shell (%)	Unstriped shell (%)
Forest / woodland	13	87
Open countryside / hedgerows	75	25

a)   Suggest a reason for this distribution of snails.   *(2 marks)*

> *More striped snails in countryside / hedgerows ✓ because the shell blends in with the striped nature of the vegetation ✓.*

b)   Which plain-coloured snail would you expect to find most of in forest / woodland areas? Explain your answer.   *(2 marks)*

> *Brown ✓. Brown camouflages more easily with the earth / forest-floor ✓.*

Snails are part of a larger group of invertebrates called gastropods. It is thought that they evolved from a primitive group of molluscs called lophophores. This group lived underwater and filter fed on smaller marine organisms using a ring of tentacles around their mouths. Snails, on the other hand, are land-dwelling and feed on plant material using a rasping tongue or radula.

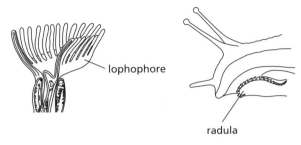

c) Explain, using ideas about natural selection, how a population of sea-dwelling lophophores could give rise to an ancestor of plant-eating gastropods in the oceans. *(4 marks)*

✓ ✓ ✓ ✓ Any four from:

mutation in lophophore population, which lacks tentacles round mouth but has more rasping mouthparts

new individual can exploit new habitats / feed off different foods

change in the environment / habitat means that there is more plant food than other sources of food; new individual's mouthparts are more adapted to feeding on plant material

they outcompete the lophophores

gastropods are more competitive

and survive to pass on mutated genes to next generation

large numbers of new individuals outcompete old population leading to the original species' extinction.

d) Suggest what may happen to the numbers of brown- and yellow-shelled snails if our climate continues to get hotter due to global warming. Give a reason for your answer. *(2 marks)*

✓ ✓ Any two from:

fewer brown-shelled / more yellow-shelled snails

brown-shelled snails overheat more easily

less competition for yellow-shelled snails.

e) Both these shell types are found within the same species. Suggest how this could have occurred. *(2 marks)*

Mutation / change in genes coding for shell colour ✓; different protein (for shell pigmentation) produced ✓.

Natural Selection

Although these ideas are much more complex than this example, a good starting point to understanding Darwin's ideas is to think of evolution as a car, in that it is the vehicle that allows us to get from A to B; or to see species X become organism Y over millions of years. Natural selection could therefore be likened to the engine in the car, in other words it is the mechanism by which the car moves. When explaining how natural selection operates, it is useful to keep in mind the natural selection 'template'. Each factor in this template was originally conceived by Charles Darwin. These factors are:

- variation
- competition
- survival of the fittest
- inheritance
- extinction.

You will gain full marks if you include all five of these factors in your answer.

Example

The picture below shows the limbs of five species of vertebrate. They are all based on the **pentadactyl** limb, which means a five-digited arm, leg, wing or flipper.

Human   Cow   Horse   Whale   Bird

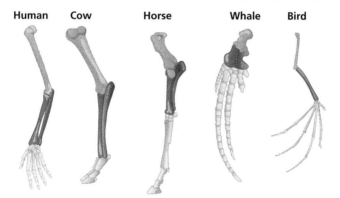

Scientists believe that the whale may have evolved from a horse-like ancestor that lived in swampy regions millions of years ago. Suggest how whales could have evolved from a horse-like mammal. In your answer, use Darwin's theory of natural selection. *(5 marks)*

✓ ✓ ✓ ✓ ✓ Any five from:

Variation – *horse-like ancestor adapted to environment / had different characteristics; named examples of different characteristics, e.g. some horse-like mammals had more flipper-like limbs (as whales have flippers); mutation in genes allowed some individuals to develop these advantageous characteristics.*

Competition *for limited resources – examples of different types of competition, e.g. obtaining food in increasingly water-logged environment.*

Survival of the fittest – *named examples of different adaptations, e.g. some horse-like mammals had more flipper-like limbs that allowed them to swim better in water than those who did not.*

Inheritance – *genes from the flipper-like individuals passed on to the next generation as they survived and the others did not.*

Extinction – *variety that is least successful does not survive and becomes extinct.*

> Take care not to word your explanation as if it was Lamarckism. Remember, the genetic change comes first – in this case genes that code for a more flipper-like pentadactyl limb. It is not the case that the horse-like ancestor recognised the need to have flippers and therefore evolved them! It may be possible for you to gain marks by explaining in more detail how speciation occurs, but this is dealt with in the next section.

Speciation

You need to be clear what a species is. Here's the working GCSE definition:

> Where two populations of a species become so different in phenotype and / or genotype that they cannot interbreed to produce fertile offspring.

Although not specifically mentioned by the exam board, it is useful to have in mind the idea of **isolation**. If two populations become isolated, e.g. by geography (flooding, rift valley formation) then, over millions or even just thousands of years they may change so much in terms of their body structures or behaviour that they cannot interbreed.

You may also be tested on the work of Alfred Wallace, who carried out much of the pioneering work on speciation.

Example

The pictures below show two species of Antelope squirrel living around the Grand Canyon in the US. They are thought to have evolved from an original population that interbred freely.

The Grand Canyon was formed from geological processes spanning millions of years, including the vertical erosion of the surrounding rocks by the Colorado River.

Harris Antelope squirrel   **White-tailed Antelope squirrel**

a)   Explain, using ideas about speciation, how these two species could have developed. *(3 marks)*

✓ ✓ ✓ Any three from:

geographical separation of original population / isolation
by Grand Canyon / geological processes / erosion of river
evolution of separate populations by natural selection
two populations evolve to the point where, if they were to intermingle again, they could not interbreed
*to produce offspring which were **fertile**.*

Alfred Wallace lived at the same time as Charles Darwin and also wrote papers on the idea of natural selection. In 1858 he wrote the following:

I ask myself: how and why do species change, and why do they change into new and well-defined species. Why and how do they become so exactly adapted to distinct modes of life; and why do all the intermediate grades die out?

b) **i)** Wallace had no knowledge of genes or DNA structure. Knowing what we know now, how would you answer his question about how organisms become 'exactly adapted'? *(2 marks)*

> ✓ ✓ Any two from:
>
> *changes in DNA / genes; mutation*
>
> *production of new proteins*
>
> *proteins determine structural or behavioural adaptations.*

ii) There are no intermediate types of Antelope squirrel. Suggest an explanation for this. *(1 mark)*

> *Intermediate types could not compete / became extinct* ✓.

Selective Breeding

Natural selection is often easier to understand if you first cover selective breeding. It seems a straightforward thing to appreciate that humans have chosen characteristics they want in both plants and animals, then bred them to achieve these characteristics. With natural selection, it is simply that nature or the principle of 'survival of the fittest' provides the driving force rather than human choice.

When describing artificial selection or selective breeding, keep in mind the general template:

- humans select the desired characteristic in parents
- the selected individuals are encouraged to breed together (at the exclusion of other individuals) or cross-pollinate in the case of plants
- the offspring resulting from the mating can then be chosen and the whole process repeated again.

Questions ask you to apply this template and often to mention drawbacks or compare it with the process of genetic engineering (see the next section).

Example

Humans have selectively bred Golden Retrievers so that they can be used as guide dogs for the visually impaired (VI). They have calm temperaments and are easily trained.

a) Describe how a dog-breeder might consistently produce retrievers useful for these training purposes. *(3 marks)*

> ✓ ✓ ✓ Any three from:
> choose adult retrievers with good temperament and ability to follow training
> arrange for these individuals to breed together
> allow puppies to grow to adulthood
> choose dogs that show the desired characteristics
> breed with other high quality dogs (from a different family)
> repeat the process many times.

b) Golden retrievers from the UK are more than twice as likely to get lymphoma as other breeds of dog. Some say this is due to the selective breeding process. Explain why this might occur and discuss whether such breeding should continue in the future. *(3 marks)*

> ✓ Any one from:
> defective alleles more likely to be paired together in close relatives
> cancer / lymphoma more likely with these genes.
> ✓ Any one from:
> cruel to limit dog's life expectancy in this way
> less variation in retrievers / reduces gene pool.
> ✓ Any one from:
> dogs with these characteristics greatly increase quality of life for VI people
> better to allow this type of selective breeding under controlled conditions
> more expensive to manage mobility for VI people by other means
> allows companionship for VI people.

Example

Give **two** examples of characteristics that farmers might want to selectively breed into their crops. *(2 marks)*

> Disease-resistant crops ✓.
>
> Large or unusual flowers ✓.
>
> (Other characteristics not included on the specification: drought-resistance; high-yield cereals.)

Genetic Engineering

Compared with selective breeding, genetic engineering is very precise and characteristics can be produced exactly in the desired organism. Genetic engineering produces genetically modified (GM) organisms – sometimes called transgenic organisms. This has led to controversy regarding the type of organisms and products produced. Questions on the drawbacks, both real and imagined, might appear in questions you have to answer.

Example

Describe how genetic engineering is different from selective breeding. *(2 marks)*

> ✓ ✓ Any two from:
>
> involves genes, not whole organisms
>
> genes transferred from one organism to another
>
> more precise in terms of passing on characteristics
>
> rapid production / process.

Example

Explain the benefit of each of these examples of genetic engineering.

a) Resistance to rose black spot fungus in roses. *(2 marks)*

> Less manually intensive / no need to remove leaves, etc. ✓.
>
> Saves money on costs / economic advantages in terms of profit ✓.

b) Introducing genes into oranges to produce larger fruit. *(1 mark)*

> Greater yield for similar input of resources from the grower ✓.

Example

Some people think that genetically engineering resistance to herbicides in plants could have harmful effects. Give **one** example of a harmful effect. *(1 mark)*

GM plants may cross-breed with wild plants, resulting in wild plants / weeds that are herbicide-resistant or that outcompete other native species ✓.

Example

a) Describe the process where artificial insulin is produced via genetic engineering. State any enzymes involved and how the final product is obtained. The first stage has been completed for you. *(4 marks)*

<u>Human gene for insulin identified.</u>

*Gene removed using **restriction enzyme** ✓.*
Bacterial plasmid 'cut open' using restriction enzyme ✓.
***Ligase enzyme** used to insert human gene into bacterial plasmid ✓.*
Insulin purified from fermenter culture and produced in commercial quantities ✓.

b) The 'cut' gene is inserted into a bacterium. Why are bacteria good host cells for the 'cut' insulin gene? *(2 marks)*

They reproduce rapidly; can be grown in large vats economically ✓.
Produce large quantities of insulin in a short time ✓.

c) What remaining step is required before the insulin can be used by patients? *(1 mark)*

Insulin needs to be purified / separated from bacteria and reactants ✓.

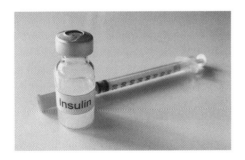

Cloning

There are four aspects of cloning technology you need to be aware of:

- cuttings
- tissue culture
- embryo transplantation
- adult cell cloning.

Cloning essentially involves the process of mitosis, although it is manipulated to a high degree for human purposes.

Below is an example of cloning plants and one about animal cloning.

Example

Describe the main sequence involved in mass producing a desirable plant through tissue culture. The first step has been done for you. *(3 marks)*

<u>Select parent with desired characteristics.</u>

Scrape off small pieces of tissue into vessels, containing nutrient and hormones ✓.

Remove plantlets / clones ✓.

Repeat process ✓.

Example

Here is a picture of Dolly the sheep with one of her offspring.

a) As with all vertebrates, Dolly grew from an embryo. What type of cell division produced this growth? *(1 mark)*

Mitosis ✓

b) Unlike most vertebrates, the method by which Dolly's embryo was produced took place in the laboratory.

Describe the steps scientists took to accomplish this. *(4 marks)*

✓ ✓ ✓ ✓ Any four from:

diploid nucleus taken from adult body cell of donor sheep

nucleus inserted into egg cell with its nucleus removed

electric shock applied to cell

egg cell with inserted nucleus allowed to divide

resulting embryo implanted into uterus of surrogate mother.

c) Dolly only lived until the age of 8 years and it took over 100 attempts to produce a viable embryo. Some people say that this is too high a price to pay for this technique. State the advantages of adult cell cloning to counter this argument. *(2 marks)*

Mass produce organisms with specific characteristics beneficial to humans ✓.

✓ Accept a maximum of **one** stated example:

high meat quality in cattle / pigs / sheep; greater milk yields in dairy cattle

high quality wool in sheep, etc.

The Theory of Evolution and History of its Development

This section of the specification details the varied theories that fed into the modern synthesis of evolution.

Example

Neo-Darwinists are scientists who have accepted Darwin's theory of natural selection but have built upon it with new ideas. They have developed the theory because of new evidence that has been found.

One of Darwin's main ideas was that successful characteristics can be passed to the next generation. How might studies of inheritance have provided further evidence for this?

(1 mark)

✓ Any one from:

Mendel's work showed that characteristics could be passed on from parents to offspring

studies of genetic material, e.g. Watson and Crick, have revealed the unit of inheritance as the gene / DNA.

Example

Many years ago, members of the giraffe family had short necks. Charles Darwin had some ideas about how organisms evolved. His theory suggested that some giraffes were born with longer necks, which gave them an advantage in finding food. They were therefore more successful and were able to breed and pass on their 'long neck' to their offspring.

Lamarck was another scientist with ideas about evolution. He had a theory that some giraffes grew longer necks to reach the leaves high up on trees. These giraffes were then more successful and were able to breed and pass their 'long necks' on to their offspring.

a) Explain why Lamarck's theory is not correct. *(2 marks)*

The giraffes would have been born with long neck or short neck genes ✓.
They could not grow long necks and change their genes during their lifetime ✓.

b) Darwin also suggested that humans and apes evolved from a common ancestor.

Give **two** reasons why Darwin's theories were not accepted by some people. *(2 marks)*

> ✓ ✓ Any two from:
> *it was against their religion (God created all organisms)*
> *there was insufficient evidence at the time*
> *people did not know about genes and mechanisms of inheritance.*

c) How many years would it take for the short-necked giraffes to evolve into the modern giraffes of today? Underline the best answer. *(1 mark)*

five thousand   five hundred thousand   five million

five million ✓

Antibiotic Resistance

This section links with Chapter 3: Infection and Response. This time the emphasis is on how microorganisms such as MRSA could have evolved to become resistant to antibiotics.

Example

Antibiotics are becoming increasingly ineffective against 'superbugs' such as MRSA.

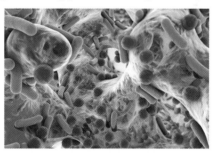

a) Use ideas about natural selection to explain how this has occurred. *(4 marks)*

> ✓ ✓ ✓ ✓ Any four from:
> *bacteria mutate – gene for resistance arises randomly*
> *non-resistant bacteria more likely to be killed by antibiotics / are less competitive*
> *antibiotic-resistant bacteria survive and reproduce often*
> *resistant bacteria pass on their genes to the next generation*
> *gene becomes more common in general population*
> *non-resistant bacteria are replaced by newer, resistant strain.*

b) Describe how medical practitioners can help reduce this problem of resistance. *(2 marks)*

Do not prescribe antibiotics for viral infections ✓.

Patients should complete the entire course of antibiotics ✓.

Evidence for Evolution

a) This diagram shows the jellyfish species, one of the first multicellular organisms to evolve.

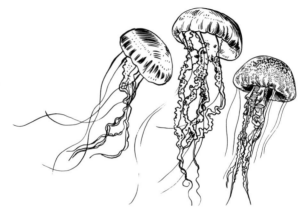

Evidence for jellyfish in the fossil record is rare. Explain why this is. *(2 marks)*

They are soft-bodies ✓.

Soft body parts decay and are therefore not preserved ✓.

b) Vertebrates are more easily fossilised. After decay of the soft parts, what stages have to occur for hard parts to be fossilised and then discovered? *(3 marks)*

Rock sediments compress organism's remains ✓.

Hard parts replaced by minerals ✓.

Tectonic movements bring remains to surface ✓.

c) How can fossils be used to identify how one form may have evolved into others? *(1 mark)*

Compare fossil froms and where they are found in the rock layers (of different ages) ✓.

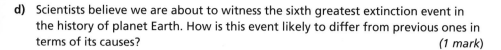

d) Scientists believe we are about to witness the sixth greatest extinction event in the history of planet Earth. How is this event likely to differ from previous ones in terms of its causes? *(1 mark)*

> *Extinction caused by the impact of human-caused climate change ✓.*

Classification

Classification simply means the sorting of things into groups, in this case, organisms.

Example

a) In the Linnaean system of classification, organisms are placed within a hierarchy of organisation. The following table shows this hierarchy with some missing levels. Complete the missing levels. *(2 marks)*

Kingdom
Class
Family
Genus
Species

> *Phylum ✓*
> *Order ✓*

b) Linnaeus also developed the **binomial system** of naming organisms. The domestic cat has the following classification:

Kingdom – Animalia

Class – Mammalia

Family – Felidae

Genus – Felis

Species – Catus

Use this information to suggest the binomial name of the common cat. *(1 mark)*

> *Felis catus ✓*

Archaeopteryx is an ancient fossilised bird. When first discovered, scientists found it hard to classify.

c) i) Using features shown in the picture, explain why *Archaeopteryx* is difficult to classify. *(2 marks)*

> Possesses features that are found in both reptiles and birds ✓.
>
> Feathers place it with birds but it also has teeth / does not have a beak like reptiles – it is an intermediate form ✓.

ii) Carl Woese developed a new classification system that divided organisms into three main groups or domains. In which group would *Archaeopteryx* be placed? *(1 mark)*

> Eukaryota ✓

iii) Scientists used to use external features to classify organisms. What technological developments have allowed more accurate systems like those of Woese to be developed? Describe how they have aided this development. *(2 marks)*

> Microscopes – to allow observation of internal features. ✓
>
> Biochemical processes – molecules common to different organisms suggest closer relationships. ✓

For more on the topics covered in this chapter, see pages 74–85 of the *Collins GCSE AQA Biology Revision Guide*.

7 Ecology

Communities – Interdependence

Ecology literally means the study of the 'homes' of organisms, although we think of the places where they live as niches, habitats, ecosystems or even the whole biosphere. Each of these terms needs to be understood in order to appreciate how organisms interact. And this is why 'interdependence' is the starting point for this chapter. It emphasises that all organisms are interconnected through relationships such as competition, predation and shared living space.

Like all the ideas covered in this biology booster guide, ecology has its basic rules and principles, but they are often more difficult to grasp because they operate on such a grand scale. We will try to pick our way through this maze and focus on the key points that examiners look for.

Example

Bird populations are a good indicator of environmental sustainability and they allow scientists to track environmental changes in particular habitats.

Scientists measured the numbers of farmland birds and woodland birds in the UK between 1972 and 2002.

Their results are shown below.

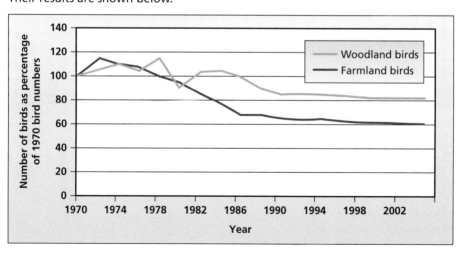

Ecology

a) Describe how numbers of farmland and woodland birds have changed between 1972 and 2002. *(4 marks)*

✓ ✓ ✓ ✓ Any four from:

overall decline in both species

initial rise in farmland birds in 1974

woodland birds show fluctuations / peaks and troughs between 1970 and 1986

rapid decline in farmland birds from 1974 onwards

more gradual decline in both species from 1986.

> Notice that you only need to describe these trends. There is no need to explain why they might have happened. That is the subject of part **c)**.

b) Suggest a reason for the overall change in numbers of farmland birds. *(1 mark)*

Farmers have cut down hedgerows and / or trees so the birds have had nowhere to nest and their food source has been reduced ✓.

c) The government wants to reverse these changes by 2020. Suggest **one** thing it could do that would help to achieve this. *(1 mark)*

✓ Any one from:

plant more trees

encourage farmers to plant hedgerows

encourage farmers to leave field edges wild as food for birds

use fewer pesticides.

d) State **two** factors that animals living in the same habitat will compete for. *(2 marks)*

✓ ✓ Any two from:

food

mates

territory.

Example

A survey of insect life was carried out on three nature reserves found in three areas of the UK.

The survey assessed the numbers of three species of insect: two butterflies (species A and B) and a parasitic wasp (species C).

Both species A and B feed on the same plant, a broad-leaved herb found in decreasing amounts on the reserves.

The parasitic wasp feeds on the caterpillars of species A and B by laying its eggs inside the live organism.

The graph below shows the proportions of each species found in the three reserves as a percentage of the total of the three populations. The 'boundary' zone is found between the northerly reserve and the southerly reserve. The average Summer temperatures for each region are also shown.

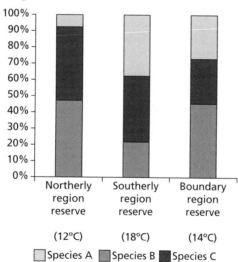

a)   Using ideas about competition and predation, compare the proportions of each species for the three regions and suggest reasons for these differences.   *(5 marks)*

✓ ✓ Any two from:

species B more numerous in north and boundary region / species A less numerous in these regions; species A more numerous in the South; species C (wasp) similar proportions in north and south; species C (wasp) less numerous in the boundary region.

✓ Any one from: **Northerly reserve** – species B outcompetes species A; species A preyed upon more by species C (wasp); lower temperatures favour species B.

✓ Any one from: **Southerly reserve** – species A outcompetes species B; species B preyed upon more by species C (wasp); higher temperatures favour species A.

✓ Any one from: **Boundary reserve** – species A is higher in number than in the northerly region as the temperature is not too cold / parasitic wasp is fewer in number; species B similar numbers compared to northerly region as not too cold; species C is fewer in number because the intermediate temperature favours **both** species A and species B.

This question might appear daunting at first, but there are many ways to pick up the marks by simply stating the patterns from the graph. Also, you are given a clue to the reasons in the stem of the question. You don't have to be more detailed about why competition and predation operates as it does in this situation. Please note that in order to gain maximum marks, you must address both the comparisons and the reasons parts of the question.

Be aware that there are many factors that affect the abundance and distribution of organisms – both **biotic** and **abiotic**. Take time to familiarise yourself with each of these categories. They include:

- biotic – availability of food, new predators, new pathogens
- abiotic – light intensity, temperature, moisture levels, soil pH and mineral content, wind intensity and direction, carbon dioxide levels for plants and oxygen levels for aquatic animals.

Adaptations in Animals

When describing adaptations it is essential to state the biological feature of the organism and **why** that feature makes it well adapted. The following questions deal with both plant and animal examples.

Example

The frilled lizard lives in Australia and feeds off cicadas, beetles, termites and mice. Its habitat exposes it to high daytime temperatures and it can often be seen basking in the morning. The frill round its neck can be quickly erected to give a startling display that can distract other animals. It can climb trees expertly and runs quickly on its hind legs.

Explain how the lizard's features make it well adapted to its environment. *(3 marks)*

> Frill can be erected to startle / scare predators / give it time to escape ✓.
>
> Running on hind legs allows it to escape predators / climbing ability allows it to escape predation ✓.
>
> Basking in the sunlight helps it to warm up for daytime activity / catching prey ✓.

Example

Scientists have discovered shrimps and giant worms that live clustered around hot vents on the ocean floor. These organisms can survive temperatures of up to 110°C. What name is given to such organisms? Underline the correct answer. *(1 mark)*

mesophiles   **gravophiles**   **extremophiles**

> extremophiles ✓

Plants

Example

Dandelions, docks and thistles are all weeds and are well adapted to compete with other plants. The drawing below shows some features of a thistle.

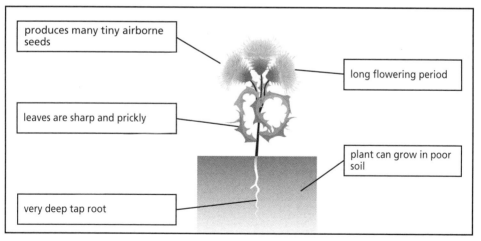

Choose any **three** of the above features. For each feature explain how it helps the thistle to compete and survive. *(3 marks)*

✓ ✓ ✓ Any three from:

many tiny airborne seeds – seeds can be carried a long way and even if many die, plenty will grow into adult plants

sharp, prickly leaves – this will deter animals from eating the thistle

very deep tap root – allows plant to get more water / makes it difficult to pull up

long flowering period – plenty of opportunity for insects to pollinate plants

plant can grow in poor soil – it can grow where other plants might not be able to.

Food Chains and Webs

At KS3 you will have learned about food chains, webs and even pyramids of numbers. At GCSE you are expected to apply these ideas further to understanding how energy flows through food chains, and how this has implications for farming and other types of food production.

Example

The food web below exists in a freshwater pond habitat.

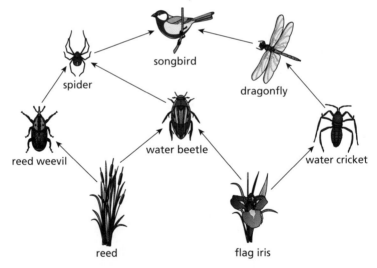

a) Write out **two** food chains; one involving three organisms, and the other involving four organisms. Each chain must include a producer, one primary consumer and one secondary consumer. The second chain must include an apex predator. *(2 marks)*

_____ ⟶ _____ ⟶ _____

_____ ⟶ _____ ⟶ _____ ⟶ _____

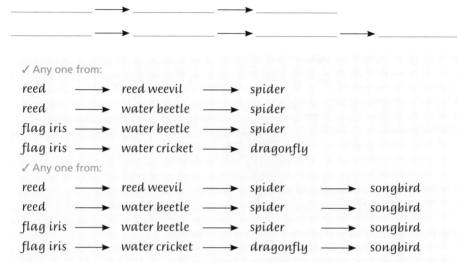

✓ Any one from:

reed ⟶ reed weevil ⟶ spider

reed ⟶ water beetle ⟶ spider

flag iris ⟶ water beetle ⟶ spider

flag iris ⟶ water cricket ⟶ dragonfly

✓ Any one from:

reed ⟶ reed weevil ⟶ spider ⟶ songbird

reed ⟶ water beetle ⟶ spider ⟶ songbird

flag iris ⟶ water beetle ⟶ spider ⟶ songbird

flag iris ⟶ water cricket ⟶ dragonfly ⟶ songbird

b) What would happen to numbers of primary consumers in the food web if numbers of water beetles declined? *(1 mark)*

They would increase ✓.

Surveying Populations

You will have carried out a required practical to measure the population size of a common species in a habitat. In all likelihood you used quadrats to measure populations of plant life.

As well as accomplishing this, you should be aware that ecologists use a variety of methods to assess distribution of species – including transects. These next questions explore these techniques.

Ecology

Example

Lichens are organisms that are sensitive to sulfur dioxide pollution.

Scientists wanted to investigate levels of pollution around an industrial area, so they carried out two line transects, as shown. At 200-metre intervals along each transect they counted the number of lichens growing on the nearest tree, as shown in the diagram.

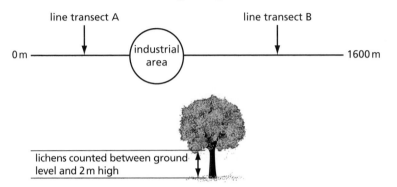

The number of lichens found are shown in the table.

Transect	200 m	400 m	600 m	800 m	1000 m	1200 m	1400 m	1600 m
A	0	0	3	4	5	8	9	8
B	2	5	8	9	9	7	8	9

a) Suggest why the results for the two transects are different. *(1 mark)*

✓ Any one from:

the direction of the wind (affecting exposure to pollutants from industrial area)

transect A may have passed over more roads, which could have caused increased pollution / sulfur dioxide levels.

b) What conclusion could you draw from the results of line transect A? *(1 mark)*

The industrial area was producing sulfur dioxide pollution ✓.

c) Suggest why scientists did not count the number of lichens on the whole tree. *(2 marks)*

The trees would have been different heights ✓.
This would affect the reliability of the data ✓.

d) Suggest **one** factor scientists were unable to control that could affect the reliability of the results. *(1 mark)*

✓ Any one from:

width of tree trunks

distance of nearest tree to transect

type of tree.

This might be an example of a question where you are asked to apply knowledge you have acquired while carrying out your required ecology practical. (The specification states that you use sampling techniques to investigate the effect of a factor on the distribution of a species.)

Example

A class of students was asked to estimate the number of daisies on the school field. The field is 60 m by 90 m and has an area of 5400 m². They decided to use quadrats that were 1 m².

a) Which is the best way of using quadrats in this investigation?
Tick **one** box. *(1 mark)*

Place all the quadrats where there are lots of plants. ☐

Place all the quadrats randomly in the field. ☐

Place all the quadrats where daisies do not grow. ☐

Place all the quadrats randomly in the field. ✓

Each student collected data by using ten quadrats. The results of one student, Shaun, are shown in the table below.

Quadrat	1	2	3	4	5	6	7	8	9	10
Number of daisies	5	2	1	0	4	5	2	0	6	3

b) i) Calculate the mean number of daisies per quadrat counted by Shaun. Show clearly how you worked out your answer. *(2 marks)*

(5 + 2 + 1 + 4 + 5 + 2 + 6 + 3) ÷ 10 ✓ = 28 ÷ 10 = 2.8 ✓

ii) Calculate the median and mode of daisies per quadrat.   *(4 marks)*

> *Median – 3.5* ✓
>
> *Working – numbers arranged in order of magnitude: 1, 2, 2, 3, 4, 5, 5, 6.*
> *The middle two numbers are 3 and 4, therefore the median is 3.5* ✓.
>
> 1 mark awarded if answer is incorrect but working is shown using the correct method.
>
> *Mode – 3.5* ✓
>
> *Working – the values 2 and 5 appear most often, the mean of these two numbers is 3.5* ✓.
>
> 1 mark awarded if answer is incorrect but working is shown using the correct method.

c) Another student, Bethany, calculated a mean of 2.3 daisies per quadrat from her results. Using Bethany's results, estimate the total number of daisies in the whole field by using the equation below. Show clearly how you work out your answer.

(2 marks)

estimated number of daisies in the field =
mean number of daisies per quadrat × number of quadrats that would fit into a field

> 2.3 × 5400 ✓ = 12 420 ✓
>
> 2 marks for a correct answer but, if incorrect, 1 mark will be given for correct working.

Predator-Prey Cycles

We have considered competition as a major factor influencing the distribution of organisms. Predation is the other main focus. The interaction between predators and prey is a popular area to be tested by examiners. Learn the basic principles, then apply these to the specific situation presented.

Example

The following information on the population of stoats and rabbits in a particular area was obtained over a period of ten years.

Year	1989	1990	1991	1992	1993	1994	1995	1996	1997	1998
No. of stoats	14	8	8	10	12	16	14	6	8	12
No. of rabbits	320	360	450	600	580	410	300	340	450	500

a) Plot these results onto the graph paper provided. *(3 marks)*

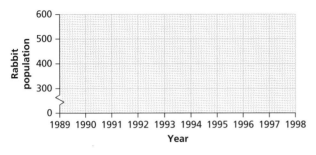

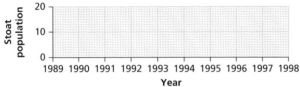

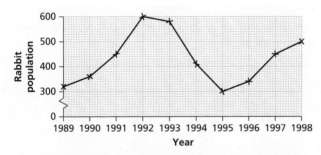

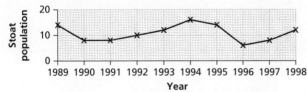

✓ ✓ for correct plotting ✓ for joining of points

b) Explain the reason for the variation in sizes of the stoat and rabbit populations.

(3 marks)

As the rabbit population decreases, there is less food (the rabbits) available for the stoats ✓.

So the stoat population decreases ✓.

Fewer rabbits are eaten, so the rabbit population increases ✓.

Ecology

You will notice in this graph that the points are connected by straight lines. This may be at odds with what you are taught in physics where a smooth curve is preferred. This method is used here because we cannot be certain of the values that lie between the points (it is not a direct relationship between numbers and time). In many physics experiments (and indeed chemistry) this is not the case and the values between the points are predictable. You could learn the answer to part **b)** by rote as it can be applied to any example of predator–prey cycles.

The Carbon Cycle and Decomposition

As well as the water cycle, you will need to understand how the element carbon moves through the living and non-living parts of the environment. You may wonder why carbon is special in this respect. As you will learn, the element forms the basis of every organic compound and therefore every biochemical process in living organisms. The main processes to look out for are respiration, photosynthesis and combustion. Decomposition is essentially a special case of respiration.

Example

Carbon is recycled in the environment in a process called the carbon cycle. The main processes of the carbon cycle are shown here.

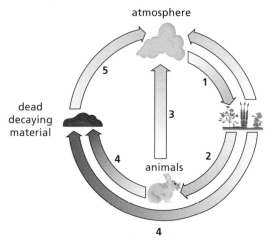

a)   Name the process that occurs at stage 3 in the diagram.   *(1 mark)*

Respiration ✓

b) The UK government is planning to use fewer fossil-fuel-burning power stations in the future. How might this affect the carbon cycle? Use ideas about **combustion** and **fossil fuel** formation in your answer. *(2 marks)*

> ✓ ✓ Any two from:
>
> *fossil fuels represent a carbon 'sink' / they absorbed great quantities of carbon many millions of years ago from the atmosphere*
>
> *combustion in power stations returns this carbon dioxide*
>
> *less burning of fossil fuels cuts down on carbon emissions*
>
> *alternative sources of energy may not add as much carbon dioxide to the atmosphere.*

As one of your required practicals you will have investigated the rate of decay in milk. Here is an example of a question where the ideas you have learned can be applied.

Example

A group of students wanted to investigate factors affecting decay. They mixed soil with small discs cut from leaves. They divided the leaf disc / soil mixture equally into four test tubes, as shown below.

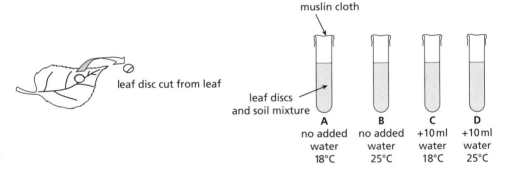

muslin cloth

leaf disc cut from leaf

leaf discs and soil mixture

A	B	C	D
no added water	no added water	+10 ml water	+10 ml water
18°C	25°C	18°C	25°C

a) In which tube would you expect the leaf discs to decay fastest? Give a reason for your answer. *(3 marks)*

> *Tube D* ✓
>
> *Because it is warm and moist* ✓*.*
>
> *And these conditions are needed for decay to occur* ✓*.*

b) The students did not add any microorganisms to the test tubes. Where will the microorganisms that cause decay come from? *(1 mark)*

> *The soil / air / surface of the leaf* ✓*.*

c) Why did the students seal the tubes with muslin cloth instead of a rubber bung?

(2 marks)

So that air / oxygen can get in ✓.

Oxygen is required for decomposition / aerobic respiration of bacteria ✓.

d) Suggest **one** way in which the students could use the leaf discs to measure the rate of decay.

(1 mark)

✓ Any one from:

they could count the number of whole discs left at the end

they could record what fraction / percentage of leaf discs decayed and find an average

they could measure the percentage decrease in the mass of the discs by measuring mass before and after the time in the soil.

When answering questions about decay, be aware that the process requires oxygen (for aerobic respiration of the bacteria). The bacteria also require moisture and warmth (don't say heat – this kills bacteria!).

Applications of Decomposition

Two applications are covered in the AQA specification: compost and biogas. There is a question on each here.

Example

Jim, Robert and Harriet are all keen gardeners. They make their own compost by collecting leaves from the garden and putting them in a heap. Jim and Robert cover their heaps with thick black plastic to absorb the heat from the Sun. Their compost heaps are shown below.

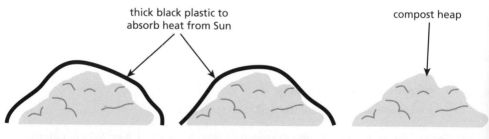

thick black plastic to absorb heat from Sun

compost heap

Jim mixes his compost by turning it over with a spade every week

Robert leaves his compost heap untouched for six months

Harriet mixes her compost occasionally

a) Who is most likely to produce useable compost first? *(1 mark)*

Jim ✓

b) Give a reason for your answer in **a)**. *(2 marks)*

Jim's compost is warmest and bacteria / enzymes work best at warmer temperatures ✓. Turning over ensures that more of the compost is exposed to more oxygen / bacteria ✓.

c) Suggest **two** environmental reasons why it is good to recycle organic waste through composting. *(2 marks)*

✓ ✓ Any two from:

if organic waste is sent to landfill, it will produce methane, which is a greenhouse gas

recycling the waste means less space is taken up in landfill sites

people will not need to buy peat-based compost (destruction of peat bogs is an environmental issue).

d) Suggest **one** economic reason why someone might want to make their own compost. *(1 mark)*

The compost can be used on plants, avoiding the need to buy compost or fertilisers ✓.

Example

The diagram shows a digester that can make use of human sewage. The gas produced from this vessel contains a mixture of 60% methane and 40% carbon dioxide.

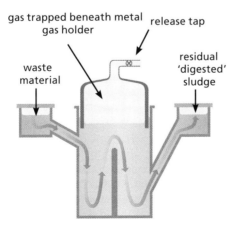

gas trapped beneath metal gas holder

release tap

waste material

residual 'digested' sludge

Ecology

a) What is the name of this type of biofuel? *(1 mark)*

Biogas ✓

b) Give **two** uses of this biofuel. *(2 marks)*

✓ ✓ Any two from:
burned to generate electricity
burned to produce hot water / steam / for central heating
used as fuel in houses
used for cooking.

c) Explain why these types of digesters are useful in remote parts of the world. *(1 mark)*

Remote areas don't have access to mains electricity; burning biogas can generate electricity ✓.

> Do not confuse biogas with biofuel. Biofuels are petrol or diesel additives produced from plant sources.

Species Distribution

Distribution of species can be studied on a macro scale, covering whole continents, or can operate on a smaller scale where species occupy niches at different times depending on factors such as temperature, water availability and composition of atmospheric gases (especially oxygen).

Example

Scientists studied a square kilometre of desert characterised by sand dunes. The graph shows how the numbers of two different species of reptile changed during the course of a 12-hour period. It also shows the air temperature. Scientists counted the animals by laying traps for them, then releasing them after counting. They marked the animals they released.

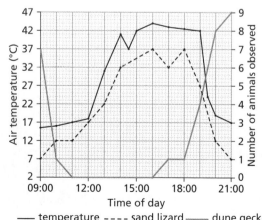

— temperature - - - - sand lizard —— dune gecko

a) Describe the trends in the numbers of the two reptiles during daylight and night time. *(3 marks)*

✓ ✓ ✓ Any three from:

sand lizard numbers increase with temperature

quote data, e.g. peak sand lizards between 16:00 and 18:00 / lowest numbers at 09:00 and 21:00

dune gecko numbers decrease as temperature increases

quote data, e.g. peak number at 21:00 / lowest or zero numbers found between 11:00 and 16:00.

> Note that credit is given for quoting data. This is always a good thing to do when numbers are provided on the axis and there is more than one mark being awarded. Make sure that you are clear and concise about the numbers quoted and that they are significant. For example, it is not particularly noteworthy that four dune geckos appeared at 19:00.

b) Suggest a reason for the differences in these trends. *(1 mark)*

✓ Any one from:

sand lizard is better adapted to higher temperatures / dune gecko is better adapted to lower temperatures

two lizards occupy different niches / avoid competition with each other for the same food sources.

Biodiversity and Conservation

Most questions link the ideas of biodiversity and conservation. The main ideas are:

- biodiversity is a measure of the number of different species found in a given area or region
- high biodiversity is desirable from an ecological point of view
- adverse human impact reduces biodiversity
- carefully applied conservation measures can maintain biodiversity.

Example

Organic foods have become popular in recent years. They are grown without the use of pesticides and fertilisers.

A government report in 2007 showed that the production of some organic foods is more damaging to the environment than their non-organic equivalents.

However, supporters of organic farming claim that it is better than non-organic farming in conserving biodiversity and is better for the soil.

a) What is meant by the term biodiversity? *(1 mark)*

The range of species in a habitat ✓.

b) Why is it important to conserve biodiversity? *(1 mark)*

✓ Any one from:

they may have future uses

moral duty to maintain biodiversity

to maintain food webs and interactions between organisms in an environment.

c) State **two** features of non-organic farming that are thought to be damaging to the environment. *(2 marks)*

✓ ✓ Any two from:

application of pesticides

application of inorganic fertilisers

destruction of hedgerows

use of machinery

less humane enclosure of animals.

Example

In Ireland, four species of bumble bee are now endangered. Scientists are worried that numbers may become so low that they are inadequate to provide pollination to certain plants.

a) Explain how the disappearance of these four species of bumble bee might affect biodiversity as a whole. *(3 marks)*

Certain plant species / crops become scarce ✓.

Due to lack of pollinators ✓.

Other species that depend on these plants will be endangered / reduced in number; biodiversity falls ✓.

b) Threats to bumble bees are many: the possible effect of pesticides such as neonicotinoids, loss of suitable habitat, disappearance of wildflowers, new diseases.

Suggest **two** conservation measures that might help increase numbers of bumble bees in the wild. *(2 marks)*

✓ ✓ Any two from:

reduce usage of pesticides such as neonicotinoids

plant species that attract bumble bees (flowers, etc.)

farm bees in hives and release into the wild

disease prevention measures to act against foul-brood disease / isolate colonies / remove diseased colonies.

Waste Management

The AQA specification requires that you are able to describe the effects of specific pollution problems: sewage, fertilisers, toxic chemicals, smoke and acidic gases in the air, landfill. The next question is in context of acid rain and its effects on plants.

Example

Dylan and Molly investigated the effect of sulfur dioxide on the germination of cress seeds. The diagram shows their apparatus.

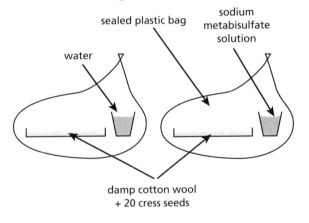

sealed plastic bag

water

sodium metabisulfate solution

damp cotton wool + 20 cress seeds

- Sodium metabisulfate solution gives off sulfur dioxide.

- Both bags were left in a warm laboratory for five days.

a) What was the independent variable in the investigation? *(1 mark)*

  The presence of water or sodium metabisulfate ✓.

b) Suggest the main reason for using sealed plastic bags. *(1 mark)*

  To keep the gases in ✓.

Ecology

Dylan and Molly counted the number of seeds that had germinated after five days. Their results are shown in the table below.

	Number of germinated seeds
Water	18
Sodium metabisulfate	12

c) i) What conclusion can Dylan and Molly draw from their results? *(1 mark)*

That sodium metabisulfate affects (decreases) the germination of cress seeds ✓.

ii) Are their results reliable? Explain your answer. *(2 marks)*

No ✓. They need to repeat the experiment to increase the reliability of their results ✓.

d) Suggest how acid rain might affect the biodiversity of plants. *(2 marks)*

Plant biodiversity lowered (no marks) *due to plants being killed through*
✓ ✓ Any two from:
effects of acid on bark
leaves being bleached / damaged
soil nutrients being affected
pH of soil being lowered.

Land Use

Example

The picture shows a peat bog.

Peat is formed by the compression of plant remains over millions of years.

Peat bogs are ancient habitats that can act as carbon sinks, which means they can store and 'lock up' carbon.

a) Explain how a peat bog acts in this way. In your answer use ideas about the carbon cycle and photosynthesis. *(3 marks)*

✓ ✓ ✓ Any three from:

photosynthesis absorbs carbon dioxide from the atmosphere

plants die but do not decay

fossilisation of plant remains

peat containing carbon compounds remains in the ground.

b) Peat bogs are declining in area across the world due to extraction for gardening purposes. Explain the effects of this habitat destruction on:

- biodiversity
- climate change. *(3 marks)*

Biodiversity lowered because habitats are destroyed for a wide range of organisms that rely on them ✓. Extraction of peat releases carbon dioxide ✓, which increases global temperatures / hastens climate change ✓.

Energy Transfer in Ecosystems

Example

a) The following organisms are found in a rock pool on a beach: seaweed (producer / plant), sea snails (herbivores), starfish (carnivore). Using this information, sketch a likely pyramid of biomass for these organisms. *(2 marks)*

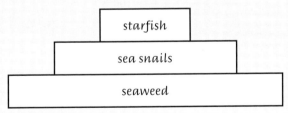

1 mark for correct sequence of organisms ✓
1 mark for correct shape of pyramid ✓

b) Explain why the pyramid has the shape that it does. *(3 marks)*

Energy lost between trophic levels ✓.

Through respiration, movement of organism, body heat, undigested food ✓.

Less energy available for next trophic level ✓.

Example

The picture shows energy transfers in a cow.

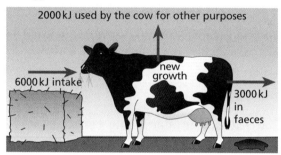

2000 kJ used by the cow for other purposes

new growth

6000 kJ intake

3000 kJ in faeces

a) What type of energy does the cow's intake consist of? Tick **one** box. *(1 mark)*

Chemical energy ☐   Light energy ☐

Potential energy ☐   Heat energy ☐

Chemical energy ✓

b) Using the formula below, calculate the energy efficiency for the cow. Show your working. *(2 marks)*

$$\text{energy efficiency} = \frac{\text{energy used usefully for new growth}}{\text{total energy taken in}} \times 100\%$$

$\frac{1000}{6000} \times 100 = 16.7\%$ 2 marks, if answer incorrect 1 mark for $\frac{1000}{6000} \times 100$.

c) Humans do not need to eat as much biomass as cows do. Explain why. *(2 marks)*

Cellulose in plant material requires more energy to digest ✓ so larger amounts need to be consumed ✓. Accept converse argument for humans.

d) Use your responses to **b)** and **c)** to explain why eating cereals and grains is more energy-efficient than eating meat. In your answer, explain the implications for agricultural land use if significant numbers of a population are vegetarian. *(6 marks)*

Part **d)** reveals the main point about why pyramids of biomass are shaped the way they are – energy losses. Make sure you can describe at least three different ways that organisms transfer energy to the environment.

✓ ✓ ✓ ✓ ✓ ✓ This is a model answer, which would score the full 6 marks.

Shorter food chains are more energy-efficient because they have fewer trophic levels. A vegetarian food chain may have only two levels: plants and humans, whereas a meat-eating diet will have three or more. We know that energy is lost at each trophic level as respiration, heat, excretion, egestion and movement. In this way, a consumer may lose up to 90% energy and only pass on 10% to the next consumer in the chain.

Producing food for a vegetarian population requires decreased areas of land because at least one trophic level is removed from the food chain. A field of wheat will feed many more humans than a field of cattle. This can be seen in a pyramid of biomass, where the width of the human trophic level is greater when humans are a primary consumer compared with when they are a secondary or tertiary consumer.

This is an example of a 'level of response' type question. It is not marked in the usual way. Rather, the examiner will weigh up the scientific points you have made and how coherently you have presented them. This is why the answer is given in the form above instead of a series of marking points. Generally speaking, the levels are awarded as follows:

Level 3: A clear, logical and coherent answer, with no wasted words or excess detail. The student understands the process and links this to reasons for any appropriate experimental approaches. **5–6 marks**

Level 2: A partial answer with errors and ineffective reasoning or linkage. **3–4 marks**

Level 1: One or two relevant points but little linkage of points or logical reasoning. **1–2 marks**

Food Security

Food security is about ensuring that with an increasing world population, all people have access to a suitable diet. There are many factors that affect this, so expect to comment on reasons for food being scarce as well as methods being used to address the problem, including GM foods. We can only touch on some of these issues here.

Example

A farmer breeds cattle for beef. The animals are kept in enclosures inside a barn so that the temperature can be regulated and they cannot move around too much.

a)   In terms of food production, explain why the farmer would want to:

i)   regulate the temperature of the environment in which the cows are kept.

(1 mark)

So energy is not wasted by the cows generating heat to keep warm ✓.

ii) restrict how much the cattle can move around. *(1 mark)*

> *So energy is not transferred to the environment through the movement of the cows ✓.*

b) Suggest **one** other **advantage** of keeping the cows inside a barn like this. *(1 mark)*

> ✓ Any suitable answer, e.g.
> *take up less space.*

c) Suggest **one disadvantage** of keeping the cows inside a barn like this. *(1 mark)*

> ✓ Any suitable answer, e.g.
> *expensive / labour-intensive / disease could spread quickly.*

d) Some people object to livestock being raised in this way. Suggest **one** reason for this. *(1 mark)*

> *Cruel because cows cannot behave in a natural way ✓.*

Sustainable Fishing

Sustainable fishing is another popular area for questions to be set. Make sure you are aware of the reasons for fish stocks going down in our oceans, together with the methods (often ineffective) of combating the problem, e.g. setting fishing quotas. This topic is well suited to data interpretation and discussing economic factors.

Example

Some scientists studied the numbers of cod caught in cool to temperate waters in the northern hemisphere. They obtained the following estimated data, which is expressed in a graph.

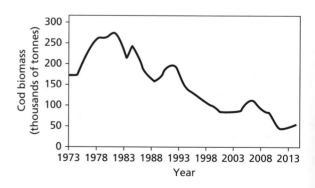

a) What was the estimated cod biomass in 1988? *(1 mark)*

160 thousand tonnes (plus or minus 10000 or in range 150–170) ✓.

b) Describe the change in estimated cod numbers between 1988 and 2013. *(2 marks)*

Overall decrease in numbers ✓.
Temporary rises between 1988–1993 and 2003–2006 ✓.

c) International fishing quotas are set in order to manage the numbers of fish in our seas.

The table shows data about fishing quotas set by an international fishing commission.

Fish species	UK quota 2013 (tonnes)	UK quota 2014 (tonnes)
Cod	11216	13123
Haddock	27507	23381
Whiting	8426	3287

i) By how much did the cod quota change between 2013 and 2014? *(1 mark)*

1907 tonnes ✓

ii) Suggest possible reasons for the decreased quota for haddock. *(2 marks)*

Numbers of haddock still declining ✓, *therefore fewer fish should be caught in order for fish stocks to recover* ✓.

d) Suggest **one** other measure that fisheries councils could take to prevent over-fishing. *(1 mark)*

✓ Any one from:

increase mesh size to allow young cod to reach breeding age
increase quotas of other fish species.

Biotechnology and Food Production

Example

a) What is mycoprotein made from? Tick **one** box. *(1 mark)*

A bacterium ☐   A virus ☐

A fungus ☐   A protozoan ☐

A fungus ✓

b) Mycoprotein is made in fermenters. What nutrient is added to help microorganisms grow and reproduce? *(1 mark)*

Glucose syrup ✓

c) The organism used in the production of mycoprotein respires. How would this type of respiration differ from the organism used in alcohol production? *(1 mark)*

Mycoprotein organism respires aerobically / yeast in alcohol production respires anaerobically ✓.

d) Genetically modified (GM) foods offer a way of producing specific varieties of crops that contain additional nutrients not present in natural varieties.

Explain how the production of golden rice can address the problem of vitamin deficiency in developing countries. *(2 marks)*

Golden rice contains an additional gene for production of beta-carotene ✓, *which is needed to make vitamin A in the body; prevents night-blindness / xerophthalmia* ✓.

Golden rice is the one example you will need to recall specific information about. However, you may be presented with information about other GM foods where you must apply your biological knowledge. Always remember that the benefits put forward for GM foods are that specific genes can be transferred from other organisms to precisely 'engineer' the qualities needed without the trial-and-error approach of selective breeding.

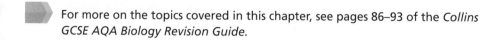

For more on the topics covered in this chapter, see pages 86–93 of the *Collins GCSE AQA Biology Revision Guide*.

Glossary

Abiotic factor – non-living component of the ecosystem

Active immunity – production of antibodies by introducing a pathogen into the body

Active site – area on an enzyme (lock) that a substrate molecule (key) can fit into

Active transport – the movement of substances against a concentration gradient; requires energy

Amino acids – building block molecules that link together to form proteins

Amylase – an enzyme that breaks down starch

Antibiotic – medicine / drug produced to combat bacterial infections

Antibody – specific protein produced in response to a specific antigen on a pathogen

Antigen – protein on the outside of a foreign cell / particle that can be recognised by antibodies

Aseptic technique – laboratory technique used in conducting experiments on microorganisms. It ensures that pathogens are not transmitted to humans, but also ensures that microorganism cultures are pure where required

Asexual reproduction – produces new individuals that are identical to their parents; does not involve the fusion of gametes

Auxins – a group of growth hormones produced in plants

Binomial system – the method of naming organisms by using their genus and species

Biotic factor – living component of the ecosystem

Carriers (genetics) – heterozygous individuals who possess a recessive gene (causing a disorder) but do not exhibit symptoms as they also have a dominant gene that 'masks' them

Cell elongation – the process of growth where cells increase in size, as opposed to cells dividing. Characteristic in tropisms

Cilia – microscopic hairs found on the surface of cells

Combustion – reaction where organic molecules break down quickly resulting in a large release of energy as light and heat

Cystic fibrosis – a genetic disorder resulting in excessive production of thick mucus and difficulty releasing digestive enzymes

Deamination – biochemical process in the liver that removes the amine group from amino acids

Denatured – where the 3-D structure of a protein is changed, usually by heat

Differentiation – process by which cells become specialised. Most cells in the embryonic state go through this process

Glossary

Dominant – an allele that always expresses itself

Double circulatory system – blood circulation divided into two – one pulmonary (to the lungs) and the other systemic (to the rest of the body)

Droplet infection – a mode of disease transmission by aerosol, e.g. coughing and sneezing to produce droplets containing pathogens

Effectors – organs (usually muscles) that respond to impulses from motor neurones – in the case of muscles, by contracting

Fertile – the ability of an organism to reproduce successfully by sexual means

Fossil fuel – coal, oil, natural gas or peat, formed from fossilisation of plant material and can be burned

Gamete – a specialised sex cell formed by meiosis

Genotype – the combination of alleles an individual has for a particular gene, e.g. BB, Bb or bb

Haploid – a chromosome set that is half in number, typically found in gametes produced by meiosis

Heterozygous – when an individual carries two different alleles for a gene, e.g. Bb

Homozygous – when an individual carries two copies of the same allele for a gene, e.g. BB or bb

Hybridoma cells – formed during the production of monoclonal antibodies by fusing cancer cells with spleen cells

Isolation (genetic) – process in speciation where populations become separated from one another

Ligase enzyme – an enzyme used in genetic engineering to 'splice' sections of DNA together

Limiting factors – in photosynthesis, where a factor, e.g. carbon dioxide concentration, limits the rate of photosynthesis despite another factor being increased

Malignant – of cancer, where cells spread from the original site of a tumour to other parts of the body

Meiosis – cell division that forms daughter cells with half the number of chromosomes of the parent cell

Memory cells – cells in the immune system that remain dormant in the body and are sensitised (can recognise) a particular pathogen if it invades

Mimicry – anti-predator strategy where an organism copies the body structure or behaviour of another to deter the predator

Mitosis – cell division that forms two daughter cells, each with the same number of chromosomes as the parent cell

Molecules – a collection of atoms joined together by bonds

Mutation – where the sequence of bases in DNA is changed and produces a new protein structure

Natural selection – process by which organisms evolve (proposed by Charles Darwin)

Non-coding base sequences – non-sense DNA – some play a role in the switching on of genes during protein production

Osmosis – the movement of water, through a partially permeable membrane, into a solution with a lower water concentration

Pathogen – a disease-causing microorganism

Pentadactyl limb – five-membered limb, e.g. wing, hand, flipper

Phenotype – the physical expression of the genotype, i.e. the characteristic shown

Photosynthesis – a process in green plants by which sunlight energy is used to synthesise carbohydrate using carbon dioxide and water

Pituitary – a small gland at the base of the brain that produces hormones; known as the 'master gland'

Plasmolysis – state in plant cells where the cytoplasm or protoplast pulls away from the cell wall due to loss of water

Polydactyly – a genetic condition caused by a dominant allele, where affected people have extra fingers or toes

Potometer – apparatus used to measure rates of transpiration in plants

Protein – large polymer molecule formed from chains of amino acids

Purified – free from contaminants. One substance only

Recessive – an allele that will only be expressed if there are two present; represented by a lower case letter

Respiration – process in all living cells whereby energy is released from the breakdown of energy-rich molecules such as glucose

Restriction enzyme – in genetic engineering, an enzyme that removes specific segments of DNA

Ribosome – sub-cellular structure where proteins are made

Statins – medicines that reduce cholesterol build-up in the body

Stent – cylindrical structure placed in arteries during surgery to maintain a wider diameter and unrestricted blood flow

Stroke volume – measurement of the volume of blood output from the heart

Transpiration – process of water loss from a plant via the leaves

Turgor pressure – in plant cells; the outward pressure on the cell wall resulting from cytoplasm and vacuole gaining or holding water

Variation – differences between individuals of the same species

Zygote – a fertilised ovum or egg cell. A zygote is also a stem cell

Notes

Notes

Acknowledgements

The authors and publisher are grateful to the copyright holders for permission to use quoted materials and images.

Every effort has been made to trace copyright holders and obtain their permission for the use of copyright material. The authors and publisher will gladly receive information enabling them to rectify any error or omission in subsequent editions. All facts are correct at time of going to press.

p99 ©2009 Jupiterimages Corporation
p115 ©Science Photo Library
All images are © Shutterstock.com and ©HarperCollins*Publishers*

Published by Collins
An imprint of HarperCollins*Publishers*
1 London Bridge Street
London SE1 9GF

ISBN: 978-0-00-827681-2

First published 2018
10 9 8 7 6 5 4 3 2 1
© HarperCollins*Publishers* Limited 2018

British Library Cataloguing in Publication Data.

A CIP record of this book is available from the British Library.

Commissioning Editor: Kerry Ferguson
Author: Tom Adams
Project Editor: Charlotte Christensen
Project Manager and Editorial: Jill Laidlaw
Cover Design: Sarah Duxbury
Inside Concept Design: Paul Oates
Text Design and Layout: QBS Learning
Production: Natalia Rebow
Printed and bound in China by RR Donnelley APS